Texts and Monographs in Symbolic Computation

A Series of the Research Institute
for Symbolic Computation,
Johannes Kepler University, Linz, Austria

More information about this series at http://www.springer.com/series/3073

Joseph Krasil'shchik • Alexander Verbovetsky
Raffaele Vitolo

The Symbolic Computation of Integrability Structures for Partial Differential Equations

Springer

Joseph Krasil'shchik
V.A. Trapeznikov Institute of Control
Sciences RAS

Independent University of Moscow
Moscow, Russia

Alexander Verbovetsky
Independent University of Moscow
Moscow, Russia

Raffaele Vitolo
Department of Mathematics
and Physics 'E. De Giorgi'
University of Salento
Lecce, Italy

ISSN 0943-853X ISSN 2197-8409 (electronic)
Texts and Monographs in Symbolic Computation
ISBN 978-3-030-10088-9 ISBN 978-3-319-71655-8 (eBook)
https://doi.org/10.1007/978-3-319-71655-8

Mathematics Subject Classification (2010): 68W30, 35Q53, 58J72, 37Kxx

Printed on acid-free paper

This Springer imprint is published by the registered company Springer International Publishing AG part of Springer Nature.
The registered company address is: Gewerbestrasse 11, 6330 Cham, Switzerland

Abstract We present a unified mathematical approach to the symbolic computation of integrability structures of partial differential equations, like Hamiltonian operators, recursion operators for symmetries and cosymmetries, and symplectic operators. The computations are carried out within the computer algebra system Reduce by the packages CDE and CDIFF.

Keywords and phrases: Symbolic computations · Integrable systems · Geometry of differential equations · Reduce · Package CDE · Package CDIFF

Preface

This book is concerned with computational aspects of the theory of integrable systems, i.e., nonlinear partial differential equations (PDEs), which admit fairly general exact solutions.

The theory of integrable systems started in 1967 with the seminal paper [39]. Since then, infinite sequences of commuting symmetries or conservation laws were observed in all integrable systems. One of the ways to generate them is to use differential operators which fulfill certain distinguished properties. These are Hamiltonian, symplectic, and recursion operators; in this book, we collectively call them *integrability structures* (IS).

Finding integrability structures and computing with them was not an easy task even for the most simple integrable equations back in the 1960s–1970s. Soon, computer algebra started to play an important role in the subject. Specialized software for computing symmetries and conserved quantities started to be publicly available since the 1980s. However, maybe due to the complexity of the subject, no specialized software for integrability structures appeared until 15 years ago.

Here is the point where this book becomes useful: it is a textbook devoted to computing integrability structures within a unified approach. The computations are carried out using the Reduce computer algebra system, which is now free software [45]. The software packages `CDE` and `CDIFF` which are used throughout the book are also freely available, being official packages of Reduce. All the examples of computations described in the book are available at the geometry of differential equations website http://gdeq.org.

The idea of this book arose after a long-standing cooperation between the authors. We hope that the reader will benefit from the unified computational viewpoint on Hamiltonian, symplectic, and recursion operators for integrable systems.

Moscow, Russia — Joseph Krasil'shchik
Moscow, Russia — Alexander Verbovetsky
Lecce, Italy — Raffaele Vitolo

Contents

Introduction

The mathematical framework of the book is that of the theory of *integrable systems* [1, 24, 26, 36, 67, 68, 83, 107, 154] viewed from the geometrical standpoint [75, 82]. According to the main ideas of the field, a partial differential equation (PDE) is integrable if it admits an infinite sequence of commuting symmetries or an infinite sequence of commuting conserved quantities [155].

It is usually very difficult to derive explicit formulas that provide all symmetries or all conserved quantities for a given PDE. Toward the end of the 1970s, the idea of bi-Hamiltonian PDEs appeared [88]. Basically, a sequence of commuting symmetries or conserved quantities can be generated by a suitable pair of differential operators of a special type called *Hamiltonian*. The first examples and theoretical facts about *recursion operators* for symmetries were also found more or less at the same time. Here a sequence of symmetries is generated by a single (pseudo-) differential operator.

Since then, it became more or less clear that finding appropriate differential operators with the needed properties for a given PDE is one of the most important steps in order to prove its integrability. In this book, all such operators will be called *integrability structures* (IS for short).

The mathematical problem of finding IS can be formulated in terms of the geometric theory of differential equations [18]. Indeed, recent papers show that finding IS is similar, from the computational viewpoint, to finding (higher or generalized) symmetries and cosymmetries of a PDE (see [82] and references therein).

In this book, we will briefly describe the mathematical theory of IS under a unifying viewpoint that allows us to treat Hamiltonian, symplectic, and recursion operator for symmetries or cosymmetries on an equal footing (see [57] or the more recent reviews [79, 82]). Moreover, we will discuss the computational problems related to IS in detail. Computing objects related to IS is an intrinsically complex problem; the number of symbols involved grows very fast. Nowadays, there are virtually no new examples that can be entirely computed by pen and paper. This is why a dedicated symbolic software is crucially important. The results of almost every recent paper in this field has been obtained using computer-aided computations. In most cases, the authors develop their own software; the choice of publicly available software for IS is quite limited (see Sect. 1.3 for a review).

Here we will describe and use the package `CDE`, a Reduce package that is designed to solve a rather wide class of computational problems that arise in the geometry of partial differential equations and the theory of integrable systems with infinite number of degrees of freedom. The package `CDE` is based on the Reduce package `CDIFF`, which was the first package to be used within the unified approach to IS [57, 79]. In order to illustrate the use of `CDE` and `CDIFF` in concrete situations, we developed several program files (see page 247 for the list). The program files are described and commented throughout the book; all of them are available at the web page of this book [80] in the "Geometry of Differential Equations" website http://gdeq.org.

We begin our exposition (Sect. 1.1) with a short introduction to the geometry of PDEs and formulate all basic problems whose solutions are discussed in the forthcoming chapters. All these problems are closely related to the integrability of PDEs. In Sect. 1.2, a short overview of distinguished software packages for computations in the field of integrable systems is presented. Chapters 2–11 contain a detailed operational exposition of the above problems and a description of the example programs that calculate the solutions of the problems. The example programs use `CDE` and `CDIFF` to write down determining equations for the problem. Then, the equations can be integrated in various ways: usually, we employ `CDIFF` to integrate algebraic systems and the package `CRACK` [149, 150] to integrate overdetermined systems of linear PDEs.

In more detail, Chap. 2 describes the procedures that introduce the main structures of an infinitely prolonged PDE $\mathcal{E}$: internal jet coordinates, total derivatives, and differential operators expressed in total derivatives (the so-called *$\mathcal{C}$-differential operators*). In particular, the procedures to compute the linearization operator $\ell_{\mathcal{E}}$ and its adjoint $\ell_{\mathcal{E}}^*$ are described. In Chap. 3, we discuss the computation of conservation laws. In addition, we discuss here how to introduce nonlocal variables associated to conservation laws and construct the corresponding differential covering [18]. Chapter 4 is devoted to computation of cosymmetries as solutions of the equation $\ell_{\mathcal{E}}^*(\psi) = 0$. The role of cosymmetries in the theory is twofold: on the one hand, they are unique characteristics of conservation laws and allow to check their nontriviality; on the other hand, they are important elements in the theory of variational symplectic (Chap. 8) and Poisson (Chap. 10) structures. In Chap. 5, the procedures to compute symmetries as solutions of the equation $\ell_{\mathcal{E}}(\varphi) = 0$ are described together with the computations of the Jacobi brackets (commutators) of symmetries.

In Chaps. 6–11, the problems that can be formulated in terms of the *tangent* and *cotangent coverings* are discussed, both in the local and nonlocal versions. The general theory of the tangent covering and the corresponding computational procedures are exposed in Chap. 6. Two important integrability invariants are related to this covering: recursion operators for symmetries and variational symplectic structures. The former are discussed in Chap. 7, where the computational procedures to compute recursion operators are given together with a description of the variational Nijenhuis bracket and the criterion for the hereditary property. Chapter 8 is devoted to the theory and computation of variational symplectic structures. The theory and computational procedures related to the cotangent covering are discussed in Chap. 9. The integrability invariants of $\mathscr{E}$ arising in this covering are variational Poisson structures and recursion operators for cosymmetries. The former constitute the topic of Chap. 10. We describe here how to compute variational bivectors on a PDE and to check their Poisson property using the variational Schouten bracket. The same bracket is also used to verify the compatibility of Poisson structures. Finally, in Chap. 11, we present computational schemes for recursion operators that act on cosymmetries.

Chapters 2–11 have a uniform structure: they all begin with a concise theoretical introduction which is illustrated by a number of examples afterward. Three examples remain the same throughout the entire exposition: the Korteweg-de Vries equation (a scalar two-dimensional evolution equation), the dispersionless Boussinesq equation (a multicomponent two-dimensional evolution system), and the Camassa-Holm equation (a scalar two-dimensional non-evolution equation). Each construction is also illustrated by multidimensional examples (the Kadomtsev-Petviashvili equation, the Plebanski equation, etc.). The theoretical parts contain no proofs but just describe the needed results. We also discuss examples in these parts to help the reader to feel how the computations should go.

In the end, we discuss some important theoretical and computational problems whose solution would lead to new mathematical understanding and/or new computational tools for IS.

The Appendix contains technical material on how to get started with Reduce and how to work with CDE for scientific computations. The Reduce programming language that is used in examples is quite simple and self-explanatory. For a deeper look into Reduce, please consult its official manual [45]. The manual contains a section that is devoted to CDE with some examples that have a partial overlap with those presented in this book.

For the readers convenience, all CDE commands (or functions) are collected in a section on page 243. The list of example programs is also available on page 247 (see the web page of this book at the Geometry of Differential Equations website [80] for the programs themselves). The list of variables which are created in each CDE run can be found on page 251.

Chapter 1
Computational Problems and Dedicated Software

Abstract In this chapter, we give an overview of the basic computational problems that arise in the study of geometrical aspects related to nonlinear partial differential equations and in the study of their integrability in particular.

We also discuss the historical development and the latest features of the Reduce software that we will use to solve the above computational problems: `CDIFF`, developed around 1990 by our colleagues P.K.H. Gragert, P.H.M. Kersten, G.F. Post, and G.H.M. Roelofs of the University of Twente and the `CDE` package, developed by one of us (RV).

Finally, we review other publicly available software that is currently used in similar computational tasks.

1.1 Computational Problems in the Geometry of PDEs and Integrability

It will not be an exaggeration to say that everyone who dealt with integrable systems or, more generally, with geometric and algebraic aspects of PDEs, sooner or later, faced the problem to compute something absolutely uncomputable by hand. Nonetheless, one is forced to compute: if not for the sake of explicit expressions (which are often huge and not very instructive) but at least to verify the existence of certain structures and/or to gain some intuition for further theoretical reasoning.

Though the arising computational problems may be (and are) different, most of them form a well-structured class whose mathematical background may be called the *calculus of total derivatives* (of $\mathscr{C}$*-differential calculus*, for short). This is not very surprising actually, because the geometry of infinitely prolonged differential equations (and, as a consequence, all their essential invariants) is based on the so-called Cartan distribution which, in turn, may be informally understood as the hull of total derivatives. For a detailed discussion, we need a deeper insight into the theory. We give below an invariant geometric exposition. All the necessary notions will be reformulated in the coordinate-wise application-oriented manner in the forthcoming sections. Our exposition is essentially based on [18, 75, 78, 82].

J. Krasil'shchik et al., *The Symbolic Computation of Integrability Structures for Partial Differential Equations*, Texts and Monographs in Symbolic Computation,
https://doi.org/10.1007/978-3-319-71655-8_1

The exposition of the theory is accompanied by discussion of the main theoretical and computational problems that naturally arise. Within the statement of each problem, we will make a reference to the chapter and section where the problem is dealt with and, in most cases, solved through `CDE` and `CDIFF` programs. The programs themselves are available to the readers at this book's web page [80] in a single zip file or in the subfolder `/packages/cde/examples`[1] of your Reduce installation.

The natural "environment" for differential equations is the *infinite jet space*. Let $\pi: \mathbb{R}^m \times \mathbb{R}^n \to \mathbb{R}^n$ be the trivial[2] bundle with $x^1, \ldots, x^n$ (independent variables) and $u^1, \ldots, u^m$ (unknown functions or dependent variables) coordinates in $\mathbb{R}^n$ and $\mathbb{R}^m$, respectively. The infinite jet space $J^\infty(\pi)$ of $\mathbb{R}^m$-valued C^∞-functions on $\mathbb{R}^n$ is the space $\mathbb{R}^\infty$ with the coordinates x^i, u^j, and u^j_σ, where $i = 1, \ldots, n$, $j = 1, \ldots, m$, and σ are a symmetric multi-index of arbitrary length $|\sigma|$ consisting of the integers $1, \ldots, n$. In particular, $u^j_\varnothing = u^j$. Denote by

$$\pi_\infty: J^\infty(\pi) \to \mathbb{R}^n$$

the natural projection to the space of independent variables. To any section $s = (u^1 = s^1(x), \ldots, u^m = s^m(x))$ (a vector function in x), we put into correspondence the section

$$j_\infty(s) = \left(u^1_\sigma = \frac{\partial^{|\sigma|} s^1}{\partial x^\sigma}, \ldots, u^m_\sigma = \frac{\partial^{|\sigma|} s^m}{\partial x^\sigma} \right) \tag{1.1}$$

which is called the *infinite jet* of s. Denote by $\Gamma(\pi)$, the space of sections of the bundle π. Then j_∞ is a map from $\Gamma(\pi)$ to $\Gamma(\pi_\infty)$. The coordinates u^j_σ in $J^\infty(\pi)$ are called *adapted* to the coordinates x^i and u^j in $M = \mathbb{R}^n$ and $E = \mathbb{R}^m \times \mathbb{R}^n$. Due to Eq. (1.1), they reflect the values of partial derivatives of sections $s: M \to E$.

For any bundle $\xi : \mathbb{R}^r \times \mathbb{R}^n \to \mathbb{R}^n$ consider its pullback

$$\begin{array}{ccc} \mathbb{R}^r \times J^\infty(\pi) & \longrightarrow & E = \mathbb{R}^r \times \mathbb{R}^n \\ {\scriptstyle \bar{\xi}} \downarrow & & \downarrow {\scriptstyle \xi} \\ J^\infty(\pi) & \xrightarrow{\pi_\infty} & M = \mathbb{R}^n. \end{array}$$

The space $\mathscr{F}(\pi;\xi)$ consists of sections of $\bar{\xi}$, i.e., of smooth vector functions $F = (F^1, \ldots, F^r)$, that may depend of arbitrary but finite number of arguments x^i, u^j_σ. In particular, $\mathscr{F}(\pi)$ denotes $\mathscr{F}(\pi;\xi)$ for one-dimensional ξ, i.e., $\mathscr{F}(\pi)$ is the algebra

---

[1] The directory names follow the Linux convention; this means that they have slashes / instead of backslashes \ like in Windows.

[2] Not necessarily trivial, actually, but we consider the simplest case here.

of smooth functions on $J^\infty(\pi)$. This algebra is filtered by the subspaces $\mathscr{F}_k(\pi)$ that consist of the functions that depend on x^i and u^j_σ with $|\sigma| \le k$. Elements of $\mathscr{F}(\pi;\xi)$ are identified with *nonlinear differential operators* acting from $\Gamma(\pi)$ to $\Gamma(\xi)$ and defined by

$$\Delta_F(s) = F \circ j_\infty(s),$$

where $s \in \Gamma(\pi)$.

The geometry of $J^\infty(\pi)$ is defined by the correspondence

$$\mathscr{C}: X = \sum_i a_i \frac{\partial}{\partial x^i} \mapsto \mathscr{C}_X = \sum_i a_i D_{x^i}, \tag{1.2}$$

where X is a vector field on M and $\mathscr{C}_X: \mathscr{F}(\pi) \to \mathscr{F}(\pi)$ is a vector field on $J^\infty(\pi)$. The operators

$$D_{x^i} = \frac{\partial}{\partial x^i} + \sum_{j,\sigma} u^j_{\sigma i} \frac{\partial}{\partial u^j_\sigma}$$

are called the *total derivatives*.[3] One has

$$[D_i, D_j] = 0$$

for all $i, j = 1, \dots, n$. The correspondence (1.2) is called the *Cartan connection*, while the space generated by the fields $\mathscr{C}_X$ is called the *Cartan distribution* on $J^\infty(\pi)$. Any differential operator expressed in terms of the total derivatives is called a *$\mathscr{C}$-differential operator*.

Example 1.1 (lifts of linear differential operators) Let

$$\Delta: (f^1, \dots, f^{r_1}) \mapsto (g^1, \dots, g^{r_2}), \qquad g^j = \sum_{\sigma,l} a^j_{\sigma,l} \frac{\partial^{|\sigma|} f^l}{\partial x^\sigma}$$

be a linear differential operator that acts from sections of a bundle $\xi_1: \mathbb{R}^{r_1} \times \mathbb{R}^n \to \mathbb{R}^n$ to those of the bundle $\xi_2: \mathbb{R}^{r_2} \times \mathbb{R}^n \to \mathbb{R}^n$. Then its *lift* acts from the sections of $\bar{\xi}_1$ to sections of $\bar{\xi}_2$ and is of the form

$$\mathscr{C}_\Delta: (F^1, \dots, F^{r_1}) \mapsto (G^1, \dots, G^{r_2}), \qquad G^j = \sum_{\sigma,l} a^j_{\sigma,l} D_\sigma(F^l),$$

[3] Everywhere below we, as a rule, use a shorter notation D_i for D_{x^i}.

where $D_\sigma = D_{i_1} \circ \cdots \circ D_{i_k}$ for $\sigma = i_1 \ldots i_k$ is a $\mathscr{C}$-differential operator from $\mathscr{F}(\pi;\xi_1)$ to $\mathscr{F}(\pi;\xi_2)$, $\operatorname{rank}\xi_i = r_i$. In particular, the total derivative D_i is the lift of the partial derivative $\partial/\partial x^i$. The lift of Δ is uniquely defined by the equality

$$\mathscr{C}_\Delta \circ j_\infty(s) = \Delta(s)$$

that must fulfill for any section $s \in \Gamma(\pi)$.

Example 1.2 (horizontal de Rham differential) Let $d\colon \Lambda^i(M) \to \Lambda^{i+1}(M)$ be the de Rham differential on $M = \mathbb{R}^n$. The lift $\mathscr{C}_d$ is denoted by

$$d_h\colon \Lambda_h^i(\pi) \to \Lambda_h^{i+1}(\pi)$$

and is called the *horizontal de Rham differential*. Elements

$$\omega = \sum_{k_1<\cdots<k_i} f_{k_1,\ldots,k_i}\, dx^{k_1} \wedge \cdots \wedge dx^{k_i}, \qquad f_{k_1,\ldots,k_i} \in \mathscr{F}(\pi)$$

are called *horizontal i-forms*, and the action of d_h is given by

$$d_h(f_{k_1,\ldots,k_i}\, dx^{k_1} \wedge \cdots \wedge dx^{k_i}) = \sum_{l=1}^{n} D_l(f_{k_1,\ldots,k_i})\, dx^l \wedge dx^{k_1} \wedge \cdots \wedge dx^{k_i}. \tag{1.3}$$

It is straightforward that d_h is a differential, i.e., $d_h \circ d_h = 0$.

Example 1.3 (linearizations) Let $P = \mathscr{F}(\pi;\xi)$, $\operatorname{rank}\pi = m$, and $\operatorname{rank}\xi = r$. Introduce the notation $\varkappa(\pi) = \mathscr{F}(\pi;\pi)$. Let $F = (F^1, \ldots, F^r) \in P$. The $\mathscr{C}$-differential operator

$$\ell_F = \begin{pmatrix} \sum_\sigma \frac{\partial F^1}{\partial u^1_\sigma} D_\sigma & \ldots & \sum_\sigma \frac{\partial F^1}{\partial u^m_\sigma} D_\sigma \\ \multicolumn{3}{c}{\dotfill} \\ \sum_\sigma \frac{\partial F^r}{\partial u^1_\sigma} D_\sigma & \ldots & \sum_\sigma \frac{\partial F^r}{\partial u^m_\sigma} D_\sigma \end{pmatrix} : \varkappa(\pi) \to P \tag{1.4}$$

is called the *linearization* of F.

Linearizations play a very important role in the theory below which leads to the following:

Problem 1.1 Given a function $F \in \mathscr{F}(\pi;\xi)$, construct its linearization operator $\ell_F\colon \varkappa(\pi) \to \mathscr{F}(\pi;\xi) = P$. The solution to this problem is given in Sect. 2.2.3.

For a $\mathscr{C}$-differential operator $\Delta\colon P_1 = \mathscr{F}(\pi;\xi_1) \to P_2 = \mathscr{F}(\pi;\xi_2)$,

$$\Delta = \begin{pmatrix} \sum_\sigma a^1_{1,\sigma} D_\sigma & \dots & \sum_\sigma a^1_{r_1,\sigma} D_\sigma \\ \dots\dots & \dots & \dots\dots \\ \sum_\sigma a^{r_2}_{1,\sigma} D_\sigma & \dots & \sum_\sigma a^{r_2}_{r_1,\sigma} D_\sigma \end{pmatrix},$$

its *adjoint* $\Delta^*: \hat{P}_2 \to \hat{P}_1$ is of the form

$$\Delta^* = \begin{pmatrix} \sum_\sigma (-1)^{|\sigma|} D_\sigma \circ a^1_{1,\sigma} & \dots & \sum_\sigma (-1)^{|\sigma|} D_\sigma \circ a^{r_2}_{1,\sigma} \\ \dots\dots & \dots & \dots\dots \\ \sum_\sigma (-1)^{|\sigma|} D_\sigma \circ a^1_{r_1,\sigma} & \dots & \sum_\sigma (-1)^{|\sigma|} D_\sigma \circ a^{r_2}_{r_1,\sigma} \end{pmatrix}, \tag{1.5}$$

where $\hat{P} = \hom_{\mathcal{F}(\pi)}(P, \Lambda^n_h(\pi))$. The operators Δ and Δ^* are related to each other by the *Green formula*

$$\langle \Delta(p_1), p_2 \rangle - \langle p_1, \Delta^*(p_2) \rangle = d_h \omega_\Delta(p_1, p_2), \tag{1.6}$$

where $p_1 \in P_1$, $p_2 \in \hat{P}_2$,

$$\langle \cdot, \cdot \rangle : P \times \hat{P} \to \Lambda^n_h(\pi)$$

is the natural pairing, while the map

$$\omega_\Delta : P_1 \times \hat{P}_2 \to \Lambda^{n-1}_h(\pi)$$

is a $\mathcal{C}$-differential operator in both arguments. Obviously, $(\Delta^*)^* = \Delta$ and $(\Delta \circ \nabla)^* = \nabla^* \circ \Delta^*$ for any $\mathcal{C}$-differential operators Δ and ∇ whenever the composition is defined

Problem 1.2 Given a $\mathcal{C}$-differential operator Δ, construct its adjoint Δ^*. The solution to this problem is given in Sect. 2.2.3.

Let $F \in P$. Consider the set

$$\mathcal{E} = \mathcal{E}_F = \{\, \theta \in J^\infty(\pi) \mid D_\sigma(F^j)(\theta) = 0, \ \forall j, \sigma \,\} \subset J^\infty(\pi).$$

This is an *(infinitely prolonged) partial differential equation* imposed on smooth vector functions $f: \mathbb{R}^n \to \mathbb{R}^m$. Let us preserve the notation $\pi_\infty: \mathcal{E} \to M$ for the restriction of the projection $\pi_\infty: J^\infty(\pi) \to M$ to $\mathcal{E} \subset J^\infty(\pi)$. A *solution* of $\mathcal{E}$ is a section s of the bundle π such that the graph of its infinite jet $j_\infty(s)$ lies in $\mathcal{E}$. Essentially, this means that we consider an equation together with all its differential (and algebraic) consequences, and these consequences are obtained by applying the total derivatives to the relations $F^j = 0$ that define the equation.

Example 1.4 Let

$$\frac{\partial u}{\partial x} = \frac{\partial v}{\partial y}, \qquad \frac{\partial u}{\partial y} = -\frac{\partial v}{\partial x}$$

be the Cauchy-Riemann system. Its infinite prolongation is given by the system of relations

$$\frac{\partial^{i+j+1}u}{\partial x^{i+1}\partial y^j} = \frac{\partial^{i+j+1}v}{\partial x^i\partial y^{j+1}}, \qquad \frac{\partial^{i+j+1}u}{\partial x^i\partial y^{j+1}} = -\frac{\partial^{i+j+1}v}{\partial x^{i+1}\partial y^j}$$

where $i, j = 0, 1, 2, \ldots$ Let us denote the corresponding adapted coordinates by $u_{i,j}$ and $v_{i,j}$, where the subscript corresponds to the $\partial^{i+j}/\partial x^i\partial y^j$-derivative. Then $\mathscr{E}$ is defined by the system

$$u_{i+1,j} = v_{i,j+1}, \qquad u_{i,j+1} = -v_{i+1,j}$$

for all $i, j = 0, 1, 2, \ldots$

We shall always[4] assume that $\mathscr{E}$ and the vector function F that defines $\mathscr{E}$ satisfy a number of conditions, the first of which is the

GENERALITY CONDITION Let $\mathscr{E}_F \subset J^\infty(\pi)$. Then

(1) The differentials $d_\theta F^1, \ldots, d_\theta F^r$ of the functions defining $\mathscr{E}$ are linearly independent at any point $\theta \in \mathscr{E}$.
(2) The natural projection $\pi_\infty\colon \mathscr{E}_F \to E = \mathbb{R}^n \times \mathbb{R}^m$ is a surjective map onto its target.
(3) The equation must be *differentially connected* which means that the only solutions of the system

$$D_1(f) = D_2(f) = \cdots = D_n(f) = 0$$

on $\mathscr{E}$ are constants.

We shall also assume the following:

REGULARITY CONDITION If a function $G \in \mathscr{F}(\pi; \xi')$ vanishes on the equation $\mathscr{E}_F$, $F \in P$, then there exists a $\mathscr{C}$-differential operator $\Delta\colon P \to F(\pi; \xi')$ such that $G = \Delta(F)$.
This means that the differential consequences of F form a complete set.

Generality condition 2 above means that differential consequences of the functions $F^1, \ldots, F^r$ do not contain "functions of order zero," i.e., functions that depend on x^i and u^j only. Condition 1 implies that, in the vicinity of any point $\theta \in \mathscr{E}$, there exists a local coordinate system such that the equations $F^1 = \cdots = F^r = 0$ may be resolved with respect to certain partial derivatives,

$$u^{j_1}_{\sigma_1} = f^1(x, u, \ldots, u^j_\sigma), \ \ldots, \ u^{j_r}_{\sigma_r} = f^r(x, u, \ldots, u^j_\sigma). \tag{1.7}$$

[4]Strictly speaking, not all these conditions are essential for *all* constructions and computations below, but we prefer not to go into unnecessary technical details.

Note that Eqs. (1.7) must be in *passive orthonomic form*, see [97]. The coordinate functions that enter the right-hand side of Eq. (1.7) are called *internal coordinates* on $\mathcal{E}$. We shall discuss these matters in more detail in Chap. 2.

Problem 1.3 Given a presentation (1.7), generate internal coordinates in $\mathcal{E}$ up to a prescribed order. The solution to this problem is given in Sect. 2.2.2.

From the definition of $\mathcal{E}$, it follows that the total derivatives, as well as any other $\mathcal{C}$-differential operator, may be restricted from $J^\infty(\pi)$ to $\mathcal{E}$, i.e., rewritten in terms of internal coordinates. In particular, this can be done for the linearization ℓ_F and its adjoint, and we use the notation $\ell_F|_{\mathcal{E}} = \ell_{\mathcal{E}}$ and $\ell_F^*|_{\mathcal{E}} = \ell_{\mathcal{E}}^*$. See Chap. 2 for a detailed discussion.

Problem 1.4 Given a $\mathcal{C}$-differential operator on $J^\infty(\pi)$, rewrite it in terms of internal coordinates. In particular, find internal coordinate presentations for the operators $\ell_{\mathcal{E}}$ and $\ell_{\mathcal{E}}^*$. The solution to this problem is given in Sect. 2.2.3.

The last condition that we impose on $\mathcal{E}$ is:

NORMALITY CONDITION For any $\mathcal{C}$-differential operator Δ, the equality $\Delta(F) = 0$ implies $\Delta|_{\mathcal{E}} = 0$.

Normality of equation $\mathcal{E}$ means that the functions $F^1, \dots, F^r$ are subject to no $\mathcal{C}$-differential relations between them. It is needed for invariance of certain constructions below.

Remark 1.1 Normality of an equation $\mathcal{E}$ is related to its cohomological properties: $\mathcal{E}$ is normal if the *compatibility complex* (see [81]) of the operator $\ell_{\mathcal{E}}$ is trivial. Equations that are not normal usually admit *gauge symmetries*. Examples of gauge invariant equations are Maxwell equations, Yang-Mills equations, Einstein equations for general relativity, and other systems arising in field theories.

For the objects restricted to $\mathcal{E}$, we shall use a notation similar to that on $J^\infty(\pi)$, e.g., $\mathcal{F}(\mathcal{E})$ will denote smooth functions on $\mathcal{E}$, $\Lambda_h^i(\mathcal{E})$ will stand for horizontal i-forms, etc. To simplify notation, when no misunderstanding arises, we shall usually denote the restrictions of the total derivatives to $\mathcal{E}$ by the same symbols D_i.

Fix an equation $\mathcal{E} \subset J^\infty(\pi)$ and a set of internal coordinates $\mathbb{I}$ on $\mathcal{E}$. A *vertical*[5] vector field

$$S = \sum_{\sigma, j \in \mathbb{I}} S_\sigma^j \frac{\partial}{\partial u_\sigma^j}, \qquad S_\sigma^j \in \mathcal{F}(\mathcal{E})$$

is a (*higher infinitesimal*) *symmetry* of $\mathcal{E}$ if

$$[D_i, S] = 0, \qquad i = 1, \dots, n.$$

[5] With respect to the projection π_∞.

Any symmetry is of the form of an *evolutionary vector field*

$$S = \mathbf{E}_\varphi = \sum_{\sigma, j \in \mathbb{I}} D_\sigma(\varphi^j) \frac{\partial}{\partial u^j_\sigma}, \tag{1.8}$$

where $\varphi = (\varphi^1, \dots, \varphi^m) \in \mathscr{F}(\mathscr{E}; \pi)$ is the *generating function* of S and satisfies the equation

$$\ell_{\mathscr{E}}(\varphi) = 0. \tag{1.9}$$

The set of solutions to Eq. (1.9) is denoted by $\mathrm{sym}(\mathscr{E})$ and forms an $\mathbb{R}$-Lie algebra with respect to to the *Jacobi bracket*

$$\{\varphi_1, \varphi_2\}^j = \sum_{\sigma, l \in \mathbb{I}} \left(D_\sigma(\varphi_1^l) \frac{\partial \varphi_2^j}{\partial u^l_\sigma} - D_\sigma(\varphi_2^l) \frac{\partial \varphi_1^j}{\partial u^l_\sigma} \right). \tag{1.10}$$

Equation (1.10) can be rewritten as

$$\{\varphi_1, \varphi_2\} = \mathbf{E}_{\varphi_1}(\varphi_2) - \mathbf{E}_{\varphi_2}(\varphi_1)$$

or, due to the identity

$$\mathbf{E}_\varphi(f) = \ell_F(\varphi), \tag{1.11}$$

as

$$\{\varphi_1, \varphi_2\} = \ell_{\varphi_2}(\varphi_1) - \ell_{\varphi_1}(\varphi_2). \tag{1.12}$$

Consequently, we have:

Problem 1.5 Given an equation $\mathscr{E}$, find solutions of Eq. (1.9) up to a prescribed order, and compute their brackets (1.10). The solution to this problem is given in Chap. 5.

Let $s = s(x^1, \dots, x^n)$ be a solution of $\mathscr{E}$ and $\varphi = (\varphi^1, \dots, \varphi^m)$ be a symmetry of this equation, where

$$\varphi^j = \varphi^j(\dots, x^i, \dots, u^\alpha_\sigma, \dots), \qquad i = 1, \dots, n, \quad j, \alpha = 1, \dots, m, \quad |\sigma| \leq k.$$

Then solutions $s_\tau = s(x^1, \dots, x^n; \tau)$ of the evolution equation

$$\frac{\partial s_\tau}{\partial \tau} = \varphi|_{j_\infty(s)} \tag{1.13}$$

satisfying the initial condition $s_0 = s$ are solutions of $\mathscr{E}$ again. Thus, Eq. (1.13) describes evolution of solutions governed by a given symmetry φ. In particular, τ-independent solutions are called *invariant* with respect to φ; φ-invariant solutions are described by a PDE that contains one less independent variable and is called the φ-reduction of $\mathscr{E}$. More generally, one can consider a Lie subalgebra $\mathfrak{g} \subset \operatorname{sym}(\mathscr{E})$ and the corresponding $\mathfrak{g}$-*reduction* of $\mathscr{E}$. This reduction will contain $n - \dim \mathfrak{g}$ independent variables.

Problem 1.6 Given an equation $\mathscr{E}$ and a Lie subalgebra $\mathfrak{g} \subset \operatorname{sym}(\mathscr{E})$, describe the $\mathfrak{g}$-reduction of $\mathscr{E}$ and find $\mathfrak{g}$-invariant solutions.

Remark 1.2 A particular class of solutions of Problem 1.6 is that of the so-called *traveling-wave solutions*, i.e., the ones that are invariant with respect to a linear combination of t and x translations.

A horizontal form[6]

$$\omega = \sum_{i=1}^{n} a_i\, dx^1 \wedge \cdots \wedge \widehat{dx^i} \wedge \cdots \wedge dx^n \in \Lambda_h^{n-1}(\mathscr{E}) \tag{1.14}$$

is a *conservation law* of $\mathscr{E}$ if $d_h\omega = 0$, i.e., if

$$\sum_{i=1}^{n} (-1)^{i-1} D_i(a_i) = 0. \tag{1.15}$$

A conservation law ω is *trivial* if ω is exact, i.e., it is of the form $\omega = d_h\theta$ for some $\theta \in \Lambda_h^{n-2}(\mathscr{E})$. Note, strictly speaking, a conservation law is the equivalence class of such a form modulo the space of exact forms.

To explain the terminology, consider an example.

Example 1.5 Let $\mathscr{E}$ be an evolution equation of the form $u_t = f(t, x, u, \dots, u_k)$, where $u_k = \partial^k u/\partial x^k$, and ω be a conservation law of this equation, i.e.,

$$\omega = A\, dx + B\, dt, \qquad D_t(A) = D_x(B).$$

Then

$$\frac{d}{dt}\int_a^b A\, dx = \int_a^b D_t(A)\, dx = \int_a^b D_x(B)\, dx = B|_{x=b} - B|_{x=a}$$

on any solution of $\mathscr{E}$. If we assume that $\lim_{x\to\pm\infty} B = 0$, then the quantity $\mathbf{A} = \int_{-\infty}^{+\infty} A\, dx$ will be independent of t on solutions.

[6] Here and below, the notation $\widehat{\cdot}$ means that the corresponding term is omitted.

On the other hand, if ω is a trivial conservation law, i.e., if

$$\omega = d_h F = D_x(F)\,dx + D_t(F)\,dt$$

then this quantity *always* vanishes under similar assumptions, independently of the equation under consideration.

Problem 1.7 Describe explicitly classes of nontrivial conservation laws up to a prescribed order. The solution to this problem is given in Chap. 3.

A direct attack to Problem 1.7 is usually inefficient; nevertheless, it can be dealt with indirectly. Namely, let ω be a conservation law of the equation $\mathcal{E} = \mathcal{E}_F$, $F \in P$, and $\bar{\omega}$ be an arbitrary extension of the form ω to the ambient space $J^\infty(\pi)$. Then, due to the regularity condition, there exists a $\mathcal{C}$-differential operator $\Delta\colon P \to \Lambda_h^n(\pi)$ such that

$$d_h\bar{\omega} = \Delta(F). \tag{1.16}$$

Then the element

$$\psi_\omega = \Delta^*(1)|_{\mathcal{E}} \in \hat{P} \tag{1.17}$$

is called the *generating function* of the conservation law ω. Its properties are described by the following:

Proposition 1.1 *Given an equation $\mathcal{E}$, generating functions of conservation laws satisfy the equation*

$$\ell_{\mathcal{E}}^*(\psi) = 0. \tag{1.18}$$

A conservation law ω is trivial if and only if $\psi_\omega = 0$.

Solutions of Eq. (1.18) are called *cosymmetries*, and their space is denoted by $\operatorname{cosym}(\mathcal{E})$.

Problem 1.8 Find cosymmetries of a given $\mathcal{E}$ up to a prescribed order, i.e., solve Eq. (1.18). The solution to this problem is given in Chap. 4.

Not every cosymmetry may be the generating function of a conservation law. If $\psi = (\psi^1, \dots, \psi^r) \in \operatorname{cosym}(\mathcal{E})$, the corresponding conservation law, if any, is found by the following procedure. Let $\bar{\psi}$ be some extension of ψ to $J^\infty(\pi) \supset \mathcal{E}$. Consider the equation

$$\sum_{i=1}^{n}(-1)^{i-1}D_i(a_i) = \sum_{j=1}^{r}\bar{\psi}^j F^j, \qquad a_i \in \mathcal{F}(\pi). \tag{1.19}$$

If a solution exists then the form

$$\omega = \sum_{i=1}^{n} a_i\,dx^1 \wedge \dots \wedge \widehat{dx^i} \wedge \dots \wedge dx^n \Big|_{\mathcal{E}}$$

is the desired conservation law.

Problem 1.9 Given a cosymmetry ψ, solve Eq. (1.19). The solution to this problem is given in Chap. 4.

Remark 1.3 In the case of evolution equations, relations between cosymmetries and conservation laws look much simpler due to the specific choice of internal coordinates. Let

$$u_t^j = f^j(t, x, \dots, u_\sigma^i, \dots), \qquad j = 1, \dots, m$$

be such an equation, where $x = (x^1, \dots, x^n)$ and u_σ^j are partial derivatives of u^j in x^i. The functions t, x^i, and u_σ^j are convenient to choose for internal coordinates on $\mathscr{E}$. Then a conservation law is a form

$$\omega = a\, dx^1 \wedge \dots \wedge dx^n + \sum_{i=1}^{n} a_i\, dt \wedge dx^1 \dots \wedge \widehat{dx^i} \wedge \dots \wedge dx^n$$

such that

$$D_t(a) = \sum_{i=1}^{n} (-1)^i D_i(a_i);$$

the coefficient a is called the *density* of ω, while $(a_1, \dots, a_n)$ is its *flux*. Then the generating function of ω is the cosymmetry

$$\psi_\omega = \left(\frac{\delta a}{\delta u^1}, \dots, \frac{\delta a}{\delta u^m} \right)$$

where

$$\frac{\delta a}{\delta u^j} = \sum_\sigma (-1)^{|\sigma|} D_\sigma \left(\frac{\partial a}{\partial u_\sigma^j} \right) \tag{1.20}$$

is the *variational derivative*. Thus, the correspondence $\omega \mapsto \psi_\omega$ is given by the *Euler operator* in the case of evolution equations. A cosymmetry ψ is a generating function if and only if its linearization is self-adjoint, i.e.,

$$\ell_\psi = \ell_\psi^*. \tag{1.21}$$

Problem 1.10 Compute variational derivatives of a given function. The solution to this problem is given in Chap. 4.

Remark 1.4 Coming back to Remark 1.3, we can also note the following. Let $G \in \mathscr{F}(\pi; \xi)$. Then the corresponding differential operator $\Delta_G \colon \Gamma(\pi) \to \Gamma(\xi)$ is an *Euler-Lagrange operator* if and only if $\ell_G = \ell_G^*$, [138] (cf. Eq. (1.21)). So, solving Problems 1.1 and 1.2, we can also easily solve the following important

Problem 1.11 Given a vector function $G \in \mathcal{F}(\pi; \xi)$, check whether the corresponding differential operator is an Euler-Lagrange operator.

Symmetries and cosymmetries of a given equation are important invariants in the study of its integrability, but they are not the only ones. The others are *recursion operators*, *Poisson*[7] and *symplectic structures*, *Lax pairs*, etc. All these invariants are closely related to the theory of *differential coverings* (i.e., to *nonlocal geometry* of PDEs).

Consider a new space $\tilde{\mathcal{E}} = \mathbb{R}^l \times \mathcal{E}$, where $l = 1, \dots, \infty$, and the projection $\tau: \tilde{\mathcal{E}} \to \mathcal{E}$. Choose coordinates w^α in $\mathbb{R}^l$. Assume that the space $\tilde{\mathcal{E}}$ is endowed with the operators

$$\tilde{D}_i = D_i + \sum_\alpha X_i^\alpha \frac{\partial}{\partial w^\alpha}$$

such that they pair-wise commute, i.e.,

$$D_i(X_j^\alpha) - D_j(X_i^\alpha) + \sum_\beta \left(X_i^\beta \frac{\partial X_j^\alpha}{\partial w^\beta} - X_j^\beta \frac{\partial X_i^\alpha}{\partial w^\beta} \right) = 0 \tag{1.22}$$

for all i, j, and α. Then we say that $\tilde{\mathcal{E}}$ (or $\tau: \tilde{\mathcal{E}} \to \mathcal{E}$) is a (*differential*) *covering* over $\mathcal{E}$, while the functions w^α are called the *nonlocal variables* in this covering. For example, *Lax pairs* are a particular case of coverings.

Problem 1.12 Find solutions of Eq. (1.22).

The problem is unsolvable in full generality, but there are efficient means to construct special types of coverings that are discussed below and widely used in various applications.

Let $\tau: \tilde{\mathcal{E}} \to \mathcal{E}$ be a covering. Then we say that a symmetry of $\tilde{\mathcal{E}}$ is a *nonlocal symmetry* of $\mathcal{E}$ in this covering. We introduce *nonlocal cosymmetries* and *nonlocal conservation laws* in a similar way. Thus, nonlocal counterparts of Problems 1.5, 1.7, and 1.8 arise.

Problem 1.13 Given a covering, compute nonlocal symmetries, cosymmetries, and conservation laws in this covering. This problem boils down to the same type of computations for generic differential equations, and it is solved in Chaps. 3, 4, and 5.

The defining equations for Problem 1.13 are as follows. Note that any $\mathcal{C}$-differential operator $\Delta = (\sum_\sigma a^j_{\sigma,i} D_\sigma)$ on $\mathcal{E}$ can be lifted to a $\mathcal{C}$-differential operator on $\tilde{\mathcal{E}}$ by

[7]Traditionally, Poisson structures on $\mathcal{E}$ are called Hamiltonian operators in the theory of integrable systems.

$$\Delta \mapsto \tilde{\Delta} = \left(\sum_{\sigma} a^j_{\sigma,i} \tilde{D}_\sigma \right).$$

In particular, this can be done with the linearization operator and its adjoint. Then any nonlocal symmetry in the covering τ is determined by vector functions $\varphi = (\varphi^1, \dots, \varphi^m)$ and $\Phi = (\Phi^1, \dots, \Phi^l)$ on $\tilde{\mathcal{E}}$ that satisfy the system of linear equations

$$\begin{aligned} \tilde{D}_i(\Phi^\alpha) &= \tilde{\ell}_{X_i^\alpha}(\varphi) + \sum\nolimits_\beta \frac{\partial X_i^\alpha}{\partial w^\beta} \Phi^\beta, \\ \tilde{\ell}_{\mathcal{E}}(\varphi) &= 0, \end{aligned} \tag{1.23}$$

where $i = 1, \dots, n$, $\alpha, \beta = 1, \dots, l$, or a bit informally, in the matrix form

$$\begin{pmatrix} \tilde{\ell}_X & \dfrac{\partial X}{\partial w} - \tilde{D}_x \\ \tilde{\ell}_{\mathcal{E}} & 0 \end{pmatrix} \begin{pmatrix} \varphi \\ \Phi \end{pmatrix} = 0.$$

Similarly, nonlocal cosymmetries are defined by the equation

$$\begin{pmatrix} \tilde{\ell}^*_X & \tilde{\ell}^*_{\mathcal{E}} \\ \left[\dfrac{\partial X}{\partial w}\right]^{\mathrm{t}} + \tilde{D}_x & 0 \end{pmatrix} \begin{pmatrix} \Psi \\ \psi \end{pmatrix} = 0,$$

where $[\cdot]^{\mathrm{t}}$ denotes the transpose of the matrix $[\cdot]$.

Solutions of the equations

$$\tilde{\ell}_{\mathcal{E}}(\varphi) = 0, \qquad \tilde{\ell}^*_{\mathcal{E}}(\psi) = 0$$

are called *shadows* of symmetries and cosymmetries in the covering $\tilde{\mathcal{E}}$, respectively.

Problem 1.14 Compute shadows of symmetries and cosymmetries in a given covering. The solution to this problem is given, respectively, in the Chaps. 5 and 4.

If a shadow φ is given and there exists a nonlocal symmetry of the form (φ, Φ), then we say that the latter is a *lift* (or reconstruction) of the shadow under consideration. Nonlocal symmetries with trivial shadows are called *invisible*. The defining equations for invisible symmetries are

$$\tilde{D}_i(\Phi^\alpha) = \sum_\beta \frac{\partial X_i^\alpha}{\partial w^\beta} \Phi^\beta, \qquad i = 1, \dots, n, \quad \alpha, \beta = 1, \dots, l, \tag{1.24}$$

cf. Eq. (1.13). Obviously, lifts are defined up to invisible symmetries. So, to Problem 1.14 we can add: given a covering and a shadow in this covering, find (if possible) lifts of this shadow (cf. [66]).

Remark 1.5 Invariantly, shadows of symmetries may be understood as derivations $\tilde{S}$ of the function algebra $\mathscr{F}(\mathscr{E})$ with values in $\mathscr{F}(\tilde{\mathscr{E}})$ such that the diagram

$$\begin{CD} \mathscr{F}(\mathscr{E}) @>{\mathscr{C}_X}>> \mathscr{F}(\mathscr{E}) \\ @V{\tilde{S}}VV @VV{\tilde{S}}V \\ \mathscr{F}(\tilde{\mathscr{E}}) @>{\tilde{\mathscr{C}}_X}>> \mathscr{F}(\tilde{\mathscr{E}}) \end{CD}$$

is commutative for any vector field X on $M = \mathbb{R}^n$.

Consider a conservation law ω and assume that the right-hand side of (1.14) contains two summands only. Then we say that ω is a *two-component conservation law*. Without loss of generality, one can assume that

$$\omega = (a_2\, dx^1 + a_1\, dx^2) \wedge dx^3 \wedge \cdots \wedge dx^n$$

with

$$D_1(a_1) = D_2(a_2).$$

Of course, all conservation laws are two-component in the two-dimensional case.

Introduce nonlocal variables w^σ, where σ is a symmetric multi-index consisting of the integers $3, \ldots, n$, and set

$$\tilde{D}_1 = D_1 + \sum_\sigma D_\sigma(a_2)\frac{\partial}{\partial w^\sigma},$$

$$\tilde{D}_2 = D_2 + \sum_\sigma D_\sigma(a_1)\frac{\partial}{\partial w^\sigma},$$

$$\tilde{D}_i = D_i + \sum_\sigma w^{\sigma i}\frac{\partial}{\partial w^\sigma}, \qquad i > 2.$$

These operators define a covering $\tau_\omega \colon \tilde{\mathscr{E}}_\omega \to \mathscr{E}$ over $\mathscr{E}$ which is called the *Abelian covering* associated to ω. It is one-dimensional if $n = 2$ and infinite-dimensional otherwise.

Problem 1.15 Construct the defining equations for Abelian coverings. The solution to this problem is given in Chap. 3.

Let an equation $\tilde{\mathscr{E}}$ cover equations $\mathscr{E}_1$ and $\mathscr{E}_2$,

$$\mathscr{B}\colon \qquad \begin{array}{ccc} & \tilde{\mathscr{E}} & \\ {\scriptstyle\tau_1}\swarrow & & \searrow{\scriptstyle\tau_2} \\ \mathscr{E}_1 & & \mathscr{E}_2; \end{array}$$

then we say that the above diagram is a *Bäcklund transformation* between $\mathcal{E}_1$ and $\mathcal{E}_2$. We also say that $\mathcal{B}$ is an auto-Bäcklund transformation when $\mathcal{E}_1 = \mathcal{E}_2$. For a detailed discussion of Bäcklund transformations, see [18, 78]; see also [23, 50, 126].

Example 1.6 Consider again Example 1.4 and set

$$w_x = v, \qquad w_y = u. \tag{1.25}$$

Then the nonlocal variable w enjoys the Laplace equation

$$\frac{\partial^2 w}{\partial x^2} + \frac{\partial^2 w}{\partial y^2} = 0.$$

On the other hand, if one sets

$$w_x = -u, \qquad w_y = v \tag{1.26}$$

then w will satisfy the Laplace equation again. Thus, Eqs. (1.25) and (1.26) define an auto-Bäcklund transformation of the Cauchy-Riemann system with $\tilde{\mathcal{E}}$ being the Laplace equation.

Bäcklund transformations and auto-Bäcklund transformations, in particular, play a very important role in the study of integrable equations, and thus we have the following:

Problem 1.16 Given two equations $\mathcal{E}_1$ and $\mathcal{E}_2$, find a Bäcklund transformation that connect them.

Remark 1.6 The last problem is quite difficult, and it is rather nontrivial to find out whether a Bäcklund transformation between two particular equations may exist. A rather complicated obstruction to existence may be found in [49], but at the moment we see no practical way to computer-based procedures that will allow to calculate this obstruction.

Another standard nonlocal construction also plays a very important role in our computational scheme. Fix an equation $\mathcal{E}$ and consider the bundle $\tau\colon W = \mathbb{R}^l \times \mathcal{E} \to \mathcal{E}$ with fiber-wise coordinates $\{w^\alpha\}$ in W. Denote by P the space of W-valued functions $\mathcal{E} \to W$ on $\mathcal{E}$. Extend W with the coordinates w^α_σ, where σ is the symmetric multi-index of the integers $1, \dots, n$ and endow the resulting space with the total derivatives

$$\tilde{D}_i = D_i + \sum_{\sigma,\alpha} w^\alpha_{\sigma i} \frac{\partial}{\partial w^\alpha_\sigma}, \qquad i = 1, \dots, n.$$

The obtained object is called the *space of horizontal jets* and is denoted by $J^\infty_h(W)$. It provides an infinite-dimensional covering $\tau_\infty\colon J^\infty_h(W) \to \mathcal{E}$ over $\mathcal{E}$.

The horizontal jet space plays the same role for differential equations in total derivatives (the so-called $\mathscr{C}$-differential equations) as "usual jets" play for equations in partial derivatives.

Let now

$$\Delta = \left(\sum_\sigma a^j_{\sigma,\alpha} D_\sigma \right) : P \to Q$$

be a $\mathscr{C}$-differential operator. Consider the $\mathscr{C}$-differential equation

$$\mathscr{E}_\Delta = \left\{ \tilde{D}_\nu \left(\sum_{\sigma,\alpha} a^j_{\sigma,\alpha} w^\alpha_\sigma \right) = 0 \right\}, \qquad |\nu| \geq 0, \quad j = 1, \dots, \operatorname{rank} Q.$$

Then the projection

$$\tau_\Delta : \mathscr{E}_\Delta \to \mathscr{E}$$

is called the Δ-*covering*. It possesses the following characteristic property:

Proposition 1.2 *Fix a $\mathscr{C}$-differential operator $\Delta : P \to Q$, and let $\Delta' : P' \to Q'$ be another $\mathscr{C}$-differential operator and $\tilde{\Delta}'$ be its natural lift to the Δ-covering. Then*:

(1) *Solutions of the equation*

$$\tilde{\Delta}'(b) = 0, \tag{1.27}$$

where $b = (b^1, \dots, b^{m'})$, $m' = \operatorname{rank} P'$, and b^j are of the form

$$b^j = \sum_{\sigma,\alpha} b^j_{\sigma,\alpha} w^\alpha_\sigma,$$

are in one-to-one correspondence with the classes of $\mathscr{C}$-differential operators

$$\nabla = \left(\sum_\sigma b^j_{\sigma,\alpha} D_\sigma \right) : P \to P'$$

such that

$$\square \circ \Delta = \Delta' \circ \nabla$$

modulo operators of the form $\nabla = \nabla' \circ \Delta$, where $\square$ is some $\mathscr{C}$-differential operator acting from Q to Q'.

(2) *Fiber-wise linear conservation laws of $\mathscr{E}_\Delta$ are in one-to-one correspondence with solutions of equation $\Delta^*(\varphi) = 0$.*

Table 1.1 Types of Δ-coverings and related IS

	Δ	Δ'	Discussed in
Case 1	$\ell_{\mathscr{E}}$	$\ell_{\mathscr{E}}$	Chapter 7
Case 2	$\ell_{\mathscr{E}}$	$\ell_{\mathscr{E}}^*$	Chapter 8
Case 3	$\ell_{\mathscr{E}}^*$	$\ell_{\mathscr{E}}$	Chapter 10
Case 4	$\ell_{\mathscr{E}}^*$	$\ell_{\mathscr{E}}^*$	Chapter 11

In particular, the operator ∇ takes $\ker \Delta$ to $\ker \Delta'$.

Problem 1.17 Given an equation $\mathscr{E}$ and a $\mathscr{C}$-differential operator Δ, construct the defining equations for the Δ-covering over $\mathscr{E}$. Examples of solution to this problem are represented by the constructions below.

We apply below this construction in the four cases presented in Table 1.1. In the case when $\Delta = \ell_{\mathscr{E}}$, the corresponding Δ-covering is called the *tangent covering* over $\mathscr{E}$ and is denoted by $\mathscr{T}\mathscr{E}$. It is a complete analog for the tangent bundle in the category of differential equations; in a similar way, the $\ell_{\mathscr{E}}^*$-covering, which we call the *cotangent covering*, is an analog of the cotangent bundle (denoted by $\mathscr{T}^*\mathscr{E}$), see a detailed discussion in [82].

If $\mathscr{E}$ is given by the relations $F = 0$, then $\mathscr{T}\mathscr{E}$ is of the form

$$F(x, \dots, u_\sigma^j, \dots) = 0, \qquad \ell_F(q) = 0,$$

while $\mathscr{T}^*\mathscr{E}$ is presented by

$$F(x, \dots, u_\sigma^j, \dots) = 0, \qquad \ell_F^*(p) = 0,$$

where $q = (q^1, \dots, q^m)$ and $p = (p^1, \dots, p^r)$ are coordinates in the fiber of the corresponding Δ-covering. It should be noted that $\mathscr{T}^*\mathscr{E}$ is always an Euler-Lagrange equation with the Lagrangian $L = \langle F, p\rangle = F^1p^1 + \dots + F^rp^r$.

Thus we have two important specifications of Problem 1.17:

Problem 1.18 Given an equation $\mathscr{E}$, construct its tangent covering $\mathscr{T}\mathscr{E}$, i.e., the Δ-covering with $\Delta = \ell_{\mathscr{E}}$. The solution to this problem is given in Chap. 6.
and

Problem 1.19 Given an equation $\mathscr{E}$, construct its cotangent covering $\mathscr{T}^*\mathscr{E}$, i.e., the Δ-covering with $\Delta = \ell_{\mathscr{E}}^*$. The solution to this problem the reader will find in Chap. 9.

Remark 1.7 The definition of the tangent covering is invariant under admissible change of coordinates and does not depend on how $\mathscr{E}$ is embedded in a jet space (a coordinate-free definition will be given in Chap. 6). As for the cotangent one, its invariance can be proved for normal equations (see [82]), and for general equations a different definition, which we do not have at the moment, seems to be needed.

Remark 1.8 It is very important to note that everywhere below we treat the tangent and cotangent coverings as supermanifolds and consider the corresponding fiber coordinates q and p (nonlocal variables) as *odd* ones. Indeed, it is natural to understand differential forms and multivectors as functions on the spaces $\mathcal{T}\mathcal{E}$ and $\mathcal{T}^*\mathcal{E}$, respectively, and both carry a natural super-algebra structure. This approach proved to be very efficient from both conceptual and computational points of view.

Let us consider the cases enlisted in Table 1.1 in more detail but recall the notation first. Let $\mathcal{E} \subset J^\infty(\pi)$ and $\pi_\infty: J^\infty(\pi) \to M$ be the natural projection to the space of independent variables. For any vector bundle η, we denote the pullback $\pi_\infty^*(\eta)$ by $\bar{\eta}$ and its restriction to $\mathcal{E}$ by $\bar{\bar{\eta}}$. Let $\mathcal{E} = \{F = 0\}$, where $F \in \Gamma(\bar{\xi})$ for some bundle ξ over M, and let $P = \Gamma(\bar{\bar{\xi}})$. Let also $\varkappa = \varkappa(\mathcal{E})$ denote the space $\Gamma(\bar{\bar{\pi}})$. Then

$$\ell_{\mathcal{E}}: \varkappa \to P, \qquad \ell_{\mathcal{E}}^*: \hat{P} \to \hat{\varkappa},$$

and we have the following particular cases of Proposition 1.2.

Case 1: The main computational problem in this case is:

Problem 1.20 Lift the operator $\ell_{\mathcal{E}}$ to the tangent covering, and solve the equation

$$\tilde{\ell}_{\mathcal{E}}(\Phi) = 0$$

for vector functions Φ linear in the variables q. The solution to this problem is given in Chap. 7.

By the first part of Proposition 1.2, the solutions to this problem will deliver us $\mathcal{C}$-differential operators $\mathcal{R}$ such that the diagram

$$\begin{array}{ccc} \varkappa & \xrightarrow{\ \ell_{\mathcal{E}}\ } & P \\ {\scriptstyle \mathcal{R}}\downarrow & & \downarrow{\scriptstyle \mathcal{R}'} \\ \varkappa & \xrightarrow{\ \ell_{\mathcal{E}}\ } & P \end{array}$$

is commutative for some $\mathcal{C}$-differential operator $\mathcal{R}'$. Operators of such a type take symmetries of $\mathcal{E}$ to symmetries and are called *recursion operators* (for symmetries). Note, by the way, that they can be understood as shadows of nonlocal symmetries in the tangent covering.

Let $\varphi = \varphi_0 \in \operatorname{sym}(\mathcal{E})$ be a symmetry (which is called a *seed symmetry*) and $\mathcal{R}$ be a recursion operator. Consider the so-called hierarchy generated by $\mathcal{R}$, i.e., the family of symmetries $\varphi_k = \mathcal{R}^k(\varphi)$. An important question from the integrability point of view is when these symmetries pair-wise commute. The answer is given in terms of the *variational Nijenhuis bracket* [74], see also [70]: if the Nijenhuis brackets of $\mathcal{R}$ with itself (which is called the *Nijenhuis tensor*, or Nijenhuis *torsion*, of $\mathcal{R}$) vanishes and $\mathcal{R}$ is invariant with respect to the seed symmetry φ (such operators are called *hereditary*), then $\{\varphi_i, \varphi_j\} = 0$ for all i and j, where $\{\cdot, \cdot\}$ is the Jacobi bracket. Thus, Problem 1.20 having been solved, we face another one:

Problem 1.21 Compute the Nijenhuis tensor of a given recursion operator $\mathcal{R}$, or, more generally, compute the Nijenhuis bracket of two recursion operators.

Case 2: The problem is as follows in this case:

Problem 1.22 Lift the operator $\ell^*_{\mathcal{E}}$ to the tangent covering, and solve the equation

$$\tilde{\ell}^*_{\mathcal{E}}(\Psi) = 0 \tag{1.28}$$

for vector functions Ψ linear in the variables q. The solution to this problem is given in Chap. 8.

From Proposition 1.2 we see that such solutions lead to $\mathcal{C}$-differential operators $\mathcal{S} = \mathcal{S}_\Psi$ such that the diagram

$$\begin{array}{ccc} \varkappa & \xrightarrow{\ell_{\mathcal{E}}} & P \\ {\scriptstyle \mathcal{S}}\downarrow & & \downarrow{\scriptstyle \mathcal{S}'} \\ \hat{P} & \xrightarrow{\ell^*_{\mathcal{E}}} & \hat{\varkappa} \end{array}$$

is commutative for some $\mathcal{C}$-differential operator $\mathcal{S}'$.

A conservation law $\omega \in \Lambda^{n-1}_h(\mathcal{E})$ is said to be *admissible* (with respect to $\mathcal{S}$) if $\psi_\omega = \mathcal{S}(\varphi)$ for some $\varphi \in \mathrm{sym}(\mathcal{E})$, where ψ_ω is the generating function of ω. For two admissible conservation laws ω_1 and ω_2, one can define their bracket with respect to $\mathcal{S}$ by

$$\{\omega_1, \omega_2\}_{\mathcal{S}} = L_{\varphi_1}(\omega_2), \tag{1.29}$$

where L_{φ_1} is the Lie derivative with respect to the evolutionary vector field $\mathbf{E}_{\varphi_2}$. Integrability of $\mathcal{E}$ is closely related to the question: When is (1.29) a *Poisson bracket*, i.e., it is skew-symmetric and satisfies the Jacobi identity? These properties of the bracket (1.29) are guaranteed by two conditions, the first being

$$(\ell^*_{\mathcal{E}} \circ \mathcal{S})^* = \ell^*_{\mathcal{E}} \circ \mathcal{S},$$

i.e., if $\mathcal{S}' = \mathcal{S}^*$, while the second one is given in terms of the variational de Rham differential δ: $\delta(\mathcal{S}) = 0$. Then $\mathcal{L}$ is called a (*variational*) *symplectic structure* on $\mathcal{E}$. Thus, the next problem is:

Problem 1.23 Given a solution Ψ of Eq. (1.28), check whether the bracket (1.29) is a Poisson bracket. The solution to this problem is given in Chap. 8.

Note that a symplectic structure on $\mathcal{E}$ can be understood as a shadow of cosymmetry in the tangent covering.

Case 3: We are in the cotangent covering now and solve the equation

$$\tilde{\ell}_{\mathcal{E}}(\Phi) = 0 \tag{1.30}$$

in this covering, i.e., look for shadows of symmetries. Then fiber-wise linear solutions give us $\mathcal{C}$-differential operators $\mathcal{P}$ such that the diagram

$$\begin{array}{ccc} \hat{P} & \xrightarrow{\ell_{\mathcal{E}}^{*}} & \hat{\varkappa} \\ {\scriptstyle \mathcal{P}}\downarrow & & \downarrow{\scriptstyle \mathcal{P}'} \\ \varkappa & \xrightarrow{\ell_{\mathcal{E}}} & P \end{array}$$

is commutative for a $\mathcal{C}$-differential operator $\mathcal{P}'$. Consequently, we have

Problem 1.24 Lift the operator $\ell_{\mathcal{E}}$ to the cotangent covering and solve Eq. (1.30). The solution to this problem is given in Chap. 10.

Similarly to the previous case, we can define the bracket

$$\{\omega_1, \omega_2\}_{\mathcal{P}} = L_{\mathcal{P}(\delta(\omega_1))}(\omega_2), \tag{1.31}$$

where δ is the Euler operator, while ω_1 and ω_2 are *arbitrary* conservation laws of $\mathcal{E}$ now and L, as before, is the Lie derivative. When is this a Poisson bracket? The answer is as follows: the bracket (1.31) is skew-symmetric and satisfies the Jacobi identity if

$$(\ell_{\mathcal{E}} \circ \mathcal{P})^* = \ell_{\mathcal{E}} \circ \mathcal{P},$$

i.e., if $\mathcal{P}' = \mathcal{P}^*$ and $[\![\mathcal{P}, \mathcal{P}]\!] = 0$, where $[\![\cdot, \cdot]\!]$ is the *variational Schouten bracket* or simply *Schouten bracket* for short.

Problem 1.25 Given a solution Φ of Eq. (1.30), check whether Equality (1.31) defines a Poisson bracket. The solution to this problem is given in Chap. 10.

The above problem has the following important generalization. Let $\mathcal{P}_1$ and $\mathcal{P}_2$ be two Poisson structures that were found when solving Problems 1.24 and 1.25. One says that they are *compatible* if $\mathcal{P}_1 + \lambda\mathcal{P}_2$ is a Poisson structure for any $\lambda \in \mathbb{R}$. One also says that such structures form a *Poisson pencil.* In the theory of integrable systems, equations that admit a Poisson pencil are called *bi-Hamiltonian.* It is known (see [88], see also [82]) that under certain conditions bi-Hamiltonian equations possess infinite hierarchies of symmetries and infinite families of conservation laws that pair-wise commute with respect to Poisson brackets defined both by $\mathcal{P}_1$ and $\mathcal{P}_2$. The procedure that generates these symmetries and conservation laws is known as the *Magri scheme*, and it is also known that two structures are compatible if and only if they commute with respect to the *Schouten bracket*. Thus, we have

Problem 1.26 Given two Poisson structures, compute their Schouten bracket. The solution to this problem is given in Chap. 10.

Of course, Problem 1.25 is a particular case of Problem 1.26.

Case 4: Being again in the cotangent covering, we solve the equation

$$\tilde{\ell}^*_{\mathscr{E}}(\Psi) = 0 \tag{1.32}$$

now, where Ψ is fiber-wise linear. Solutions lead to $\mathscr{C}$-differential operators $\bar{\mathscr{R}}$ such that the diagram

$$\begin{array}{ccc} \hat{P} & \xrightarrow{\ell^*_{\mathscr{E}}} & \hat{\varkappa} \\ {\scriptstyle \bar{\mathscr{R}}}\downarrow & & \downarrow{\scriptstyle \bar{\mathscr{R}}'} \\ \hat{P} & \xrightarrow{\ell^*_{\mathscr{E}}} & \hat{\varkappa} \end{array} \tag{1.33}$$

is commutative for some $\mathscr{C}$-differential operator $\bar{\mathscr{R}}'$. These operators take cosymmetries of $\mathscr{E}$ to cosymmetries. Let $\psi = \psi_0 \in \mathrm{cosym}(\mathscr{E})$ and $\psi_k = \bar{\mathscr{R}}^k(\psi_0)$. Assume that there exist conservation laws ω_k such that $\delta(\omega_k) = \psi_k$, where δ is the Euler operator. Then $\bar{\mathscr{R}}$ allows us to construct an infinite family of conservation laws. So, we have:

Problem 1.27 Solve Eq. (1.32), and, if possible, generate a hierarchy of conservation laws using the corresponding recursion operator $\bar{\mathscr{R}}$. The solution to this problem is given in Chap. 11.

Remark 1.9 At first glance, there exists no reasonable integrability condition for operators obtained in Case 4. Nevertheless, one can observe the following: let $\bar{\mathscr{R}}$ be such an operator. Then, due to commutativity of Diagram (1.33), one has

$$\ell^*_{\mathscr{E}} \circ \bar{\mathscr{R}} = \bar{\mathscr{R}}' \circ \ell^*_{\mathscr{E}}$$

and consequently

$$\bar{\mathscr{R}} \circ \ell_{\mathscr{E}} = \ell_{\mathscr{E}} \circ (\bar{\mathscr{R}}')^*.$$

In other words, the operator $(\bar{\mathscr{R}}')^*$ is a recursion operator for symmetries of $\mathscr{E}$. Thus, we can compute the Nijenhuis tensor of $(\bar{\mathscr{R}}')^*$, and say that $\bar{\mathscr{R}}$ is hereditary if this tensor vanishes.

Since these structures may be nonlocal (and in the case of recursion operators *are* nonlocal in most cases), the necessity to construct nonlocal odd variables arises. As it follows from general properties of Δ-coverings (see Proposition 1.2 (2)), there exists a canonical way to construct Abelian coverings over the space of tangent and cotangent coverings which we use in our computations:

Proposition 1.3 *There exists a one-to-one correspondence between symmetries of $\mathcal{E}$ and fiber-wise linear conservation laws of $\mathcal{T}^*\mathcal{E}$. Dually, cosymmetries of $\mathcal{E}$ are naturally identified with fiber-wise linear conservation laws of $\mathcal{T}\mathcal{E}$.*

We call conservation that arise on the tangent and cotangent coverings in such a way *canonical* ones. Adding the corresponding nonlocal variables to $\mathcal{E}_\Delta$, we obtain nonlocal versions of Problems 1.20–1.27:

Problem 1.28 Solve Eq. (1.27) on $\mathcal{E}_\Delta$ extended with nonlocal variables obtained from canonical conservation laws. The solution to this problem is given for the tangent and cotangent covering in the Chaps. 7, 8, 10, and 11.

Problem 1.29 Compute variational Nijenhuis bracket, variational de Rham differential, and variational Schouten bracket of the corresponding nonlocal objects.

It should be noted that we can not solve the last problem in full generality at the moment.

The above Cases 1–4 provide a unified scheme for the search of IS: symplectic operators, recursion operators for symmetries and cosymmetries, and Hamiltonian operators.

1.2 Reduce Software for the Geometry of PDEs and Integrability

In our research on the geometry of PDEs and integrable systems, we had the opportunity to use many computer algebra systems, like Mathematica, Maple, and Reduce. Within those computer algebra systems, the number of packages that are related to the geometry of PDEs is so high that it is impossible to describe all of them here. See Sect. 1.3 for a review of some software packages for computations of integrability structures.

A significant part of our research activity has been carried out in Reduce, both for tradition coming from the Twente group of symbolic computations and for more practical reasons. Indeed, in recent times (2008) Reduce became free software, and it was possible to study its source code and learn how to write optimized code for our problems. New documentation on Reduce internals has been written as an invitation to study and understand this system [106]. Moreover, several tests showed that the algebraic engine of Reduce is, at least in some cases, faster than that of many competitors. One particular situation in which Reduce performs exceptionally well is the simplification of algebraic expressions involving rational functions. This means that there can be situations in which using Reduce can be fruitful.

There are many available packages in Reduce whose tasks are in the area of geometry of PDEs. In particular there are packages for writing determining equations for symmetries [108–110, 128, 147] and conservation laws [148], for solving overdetermined systems of PDEs [149, 150], etc.

As it was already indicated in Sect. 1.1 (see also [40, 51, 57, 72, 74]), the supermanifold approach highly facilitates all the computations related to the study of IS. The advantage of this approach to IS is in the fact that many problems which would involve calculus with operators are reformulated as problems of calculus on supermanifolds. This only involves polynomial functions in odd variables, thus making symbolic computations way easier to deal with. For example:

- the problem of searching IS can be reduced to a problem whose computational complexity is the same as a generalized symmetry computation for systems of partial differential equations on supermanifolds. This means that the equations can be formulated with purely even variables and their derivatives or can be polynomials in odd variables and their derivatives;
- the problem of computing with IS (i.e., brackets of Hamiltonian operators or symplecticity tests) is reduced to the problem of computing variational derivatives of expressions which are polynomial in odd variables.

In the supermanifold approach to IS, P.H.M. Kersten was able to use existing software from the Twente research group [120, 124, 125] in order to define anticommuting coordinates and vector fields on supermanifolds and to program computations with these objects. In particular, Kersten programmed total derivatives restricted to systems of partial differential equations on supermanifolds. This was the most important step to searching IS as shadows of symmetries in specific Δ-coverings (see Problem 1.17). The effectiveness of this approach was demonstrated through a series of papers in the last 20 years [43, 54–59, 61–63, 71–74].

After that Reduce became free software, one of us (RV) packaged the old Twente software into the official Reduce package `CDIFF` , also writing a manual which is oriented to computations with IS [141]. The authors of this software package are P.K.H. Gragert, P.H.M. Kersten, G.F. Post, and G.H.M. Roelofs at the University of Twente, The Netherlands.

Later it became clear that extra work was needed in order to make the `CDIFF` package into a more effective tool in the geometry of differential equations and IS in particular. More precisely, the major drawbacks of P.H.M. Kersten's approach to `CDIFF` were:

- nonautomatic definition of even derivative symbols;
- nonautomatic definition of odd variables;
- nonautomatic definition of differential consequences of the given PDE;
- nonautomatic definition of coefficients of total derivatives.
- nonautomatic generation of ansatz for the solution of determining equations.

The above situation meant that one had to insert by hand a huge number of symbols and equations into every program, thus making the work cycle of programming a new computation very slow for all but the most basic examples.

This led one of us (RV) to writing the new Reduce package `CDE` , another official package of the Reduce distribution [142]. The new features of `CDE` are:

- There can be an arbitrary high number of independent and dependent variables; the PDEs may be systems with an arbitrary number of equations.
- Derivatives are symbols (and not procedures as in many other software packages); this follows more closely the jet space approach [18]. Derivatives up to a fixed order are automatically generated.
- Total derivatives are automatically generated up to the given order.
- Differential consequences of the (system of) PDEs up to the given order are computed. Total derivatives are automatically restricted to the given (system of) PDEs. Note that in order to achieve this result the PDE must be free of any higher integrability condition, i.e., it should be given in *passive orthonomic form* [97] (see Sect. 2.1.1).
- The above computations are automatically performed also for systems with anticommuting dependent variables.[8]
- Linearization of a vector function and its adjoint are automatically computed. These features will be extended to provide a complete calculus with $\mathscr{C}$-differential operators in the future.
- Determining equations for IS for a (system of) PDEs, even in a multi-dimensional situation, can be generated, and ansatz of various type can be automatically generated.
- Procedures for computing variational derivatives, Nijenhuis brackets of recursion operators, Schouten brackets of local Hamiltonian operators, and the variational differential for symplectic operators (again in multiple space dimensions) are included.

To the authors' knowledge, there are few other symbolic software packages that put together the abilities of searching and computing with IS (and none in Reduce). They will be discussed in Sect. 1.3. The reliability of `CDE` is now high as it has been successfully used through a number of recent papers [21, 33, 62, 79, 114, 127, 134]. Moreover, `CDE` comes with several example programs together with their results that are also provided as regression tests. This means that, if the development of new features introduces new bugs, they are usually detected by the failure in one or more examples to reproducing given results. See the Appendix for more details.

In the computer algebra part of this book, we will briefly describe the algorithms of `CDE`, and we will illustrate the mathematical problems that it can solve through examples. The parts which are shared between all examples are described only once. All examples in this book are available in the Reduce sources or at the web page of this book at the geometry of differential equations website [142].

As we already mentioned in the Introduction, we invite the interested reader to have a look at the manual of Reduce in order to learn its basic features. We stress that, for the reader's convenience, we provided in the Appendix information on installing and getting started with Reduce, as well as working with `CDE` for scientific computations.

[8]In principle, the package allows to deal with odd *independent* variables too (see, e.g., [54, 55, 59], but we do not discuss these problems here.).

1.3 Other Software for the Geometry of PDEs and Integrability

As we already said in the previous section, it is not possible to mention all software packages in the area of geometry of PDEs and/or integrability of PDEs. What we will do in this section is to describe some software packages whose primary task is to make computations with IS. We stress that none of the software described below can use odd variables or can work on super-differential equations, like the combination `CDE-CDIFF`.

Baldwin and Hereman (2010). In the paper [6] a Mathematica package,

`PDERecursionOperator.m`

for computing recursion operators is described. The package is available at the web page of one of the authors (see [6]).

The package aims at finding recursion operators for systems of polynomial evolution equations in two independent variables. Recursion operators can have one or more nonlocal terms of the type $aD^{-1} \circ b$, where a and b are functions of the field variables and their derivatives.

The algorithm relies on a weighted coordinates approach for the construction of an ansatz for the (polynomial) coefficients of the recursion operator. The equation that candidate recursion operators are required to fulfill comes from the requirement that any symmetry should be mapped to a symmetry (see Proposition 1.2). A feature that it is still missing in the package is the possibility to check the hereditary property of the recursion operator (see Chap. 7).

Barakat (2001). The Maple package `jets` [8] aims at providing a complete environment for computations on jet spaces. The facilities related to integrable systems, besides standard symmetries and conservation laws, are limited to computing the Schouten bracket of differential operators. Here, differential operators are entered as a list of coefficients. Example of computations are the list of all Hamiltonian operators that are compatible with a given Hamiltonian operator with one or two components [8]. Similar computation were done more systematically in [86].

Baran and Marvan (2010). The Maple package `Jets` [15], initially developed by Marvan [15], then also by Baran [15], is an environment for differential calculus on jet spaces and differential equations. The software has been used as a jet space environment for computations with IS in a large number of problems, although it does not deal with odd variables at the moment.

In particular, `Jets` can compute symmetries, conservation laws, zero-curvature representations, recursion operators, and many other invariants of systems of partial differential equations. Moreover, `Jets` implements the algorithms of [97]; for example, `Jets` is able to find a passive orthonomic form for a system of PDEs (if such a form exists). This feature is absent in `CDE`, despite the fact that `CDE` can only deal with systems of PDEs which are in passive orthonomic form.

The interested reader can consult the papers [9–14, 16] for computations that have been done with `Jets` on problems which are closely related to those exposed in this book.

Meshkov (2002). The Maple package `JET` [101] contains several procedures devoted to the integrability of systems of evolution PDEs in two independent variables.

Besides standard computations of symmetries and conservation laws, the package can compute recursion, symplectic (inverse Noether operators in the text of [101]), and Hamiltonian (Noether operators in the text) operators as operators that map symmetries/cosymmetries into symmetries/cosymmetries. Operators are input as lists of coefficients of D_σ (here σ just means the power of x-derivative).

The software can also can check the hereditary property, the symplectic property, and the Hamiltonian property (implectic property in the text). It can check the property of skew-adjointness. The operators are found by their definition, which yield PDEs on their coefficients. Nonlocal terms of the type $aD^{-1} \circ b$ are allowed in the operators, where a and b are functions of the field variables and their derivatives.

The package can also do some computations with Lax pairs.

Anderson (2017). The Maple package `DifferentialGeometry` [2], developed by I. Anderson, is being extended by a library of procedures which are devoted to integrable systems, named Integrable Systems Tools. At the moment of writing, we have been able to read an overview of the capabilities of the package.

The package deals with differential operators written in quite a natural language. Several operations are possible, like the composition or the commutator of operators and applying an operator to a vector function. Nonlocal operators in negative powers of the total derivatives are also allowed by the package. The package comes with a wide selection of examples on the most common integrable nonlinear PDEs.

Since the package is part of a well-established package for computations in differential geometry, it is going to be an important tool for all researchers in the field of geometry of differential equations.

Casati and Valeri (2016). Recently, an algebraic approach to the Hamiltonian formalism for integrable systems of PDEs has been started in [7]. The approach led to several new results concerning algebraic classifications of Hamiltonian operators and an alternative way to compute Schouten brackets between operators.

As it is obvious, the above algebraic theory can be translated into a computer algebra system. The new Mathematica packages `MasterPVA` and `WAlg` [22] implement some of the constructions of the algebraic theory. In particular, the Schouten bracket between local differential operators can be computed (in the algebraic language).

Chapter 2
Internal Coordinates and Total Derivatives

Abstract We describe here a general coordinate setting that allows one to deal with computational problems arising in geometry of PDEs listed in Sect. 1.1 and in the theory of integrable systems, in particular. For the convenience of reading, we expose in the beginning and in more detail the needed theoretical material that was concisely presented in Sect. 1.1. The same scheme of exposition is adopted in all the forthcoming chapters.

2.1 General Theory

Consider the Euclidean spaces $\mathbb{R}^n$ and $\mathbb{R}^m$ with coordinates $x = (x^1, \dots, x^n)$ and $u = (u^1, \dots, u^m)$, respectively, and denote by $E = J^0(n, m)$ the Cartesian product $\mathbb{R}^m \times \mathbb{R}^n$. The x coordinates play the role of independent variables, while the coordinates u stand for unknown functions (dependent variables). Let us introduce the spaces $J^k(n, m)$ with the coordinates x and u^j_σ, where $j = 1, \dots, n$ and $\sigma = i_1 \dots i_l$ is a symmetric multi-index with $0 \le l \le k$ and $1 \le i_\alpha \le n$. We use the notation $|\sigma| = l$. Then $J^k(n, m)$ is called the *jet space* of *order* k.

Remark 2.1 In particular examples, it happens to be more convenient not to enumerate dependent and independent variables but to use "individual" labels for them, such as x, y, t, u, v, etc. In this case, the jet variables will be labeled by u_x, v_{xt}, etc.

Vector functions $F = (F^1, \dots, F^r)$ on $J^k(n, m)$, $F^\alpha = F(x, u, \dots, u^j_\sigma, \dots)$, are identified with (*nonlinear*) *differential operators*:

$$\Delta_F(f^1, \dots, f^m) = \left(\dots, F^\alpha\left(x, f, \dots, \frac{\partial^{|\sigma|} f^j}{\partial x^\sigma}, \right), \dots\right).$$

J. Krasil'shchik et al., *The Symbolic Computation of Integrability Structures for Partial Differential Equations*, Texts and Monographs in Symbolic Computation,
https://doi.org/10.1007/978-3-319-71655-8_2

A *differential equation of order k* involving the unknown functions $u^1, \dots, u^m$, $u^j = u^j(x^1, \dots, x^n)$ is the set

$$\mathscr{E} = \{\, \theta \in J^k(n,m) \mid F(\theta) = 0 \,\} \subset J^k(n,m).$$

Solutions of $\mathscr{E}$ are $f = (f^1, \dots, f^m)$ such that $\Delta_F(f) = 0$.

Example 2.1 Let $m = 1$. Then in the space $J^2(n+1,1)$ one can choose the coordinates:

$$x^0, x^1, \dots, x^n, u, u_i, u_{ij}, \qquad 0 \le i \le j \le n.$$

The function

$$F = u_{00} - c \sum_{i=1}^{n} u_{ii}, \qquad c = \text{const} \in \mathbb{R}, \tag{2.1}$$

defines the linear wave equation:

$$\frac{\partial^2 u}{\partial t^2} = c\Delta_L(u),$$

where $t = x^0$ and Δ_L is the Laplace operator.

Example 2.2 Consider the spaces $\mathbb{R}^2$ with coordinates t and x and $\mathbb{R}^3$ with coordinates u, v, and w. Then the space $J^1(2,3)$ is endowed with coordinates:

$$t, x, u, v, w, u_t, u_x, v_t, v_x, w_t, w_x.$$

The first-order differential operator

$$\Delta\colon (u,v,w) \mapsto \left(\frac{\partial u}{\partial t} - w\frac{\partial w}{\partial x} - \frac{\partial v}{\partial x}, \frac{\partial v}{\partial t} + u\frac{\partial w}{\partial x} + 3w\frac{\partial u}{\partial x}, \frac{\partial w}{\partial t} - \frac{\partial u}{\partial x} \right)$$

is identified with the three-component function:

$$F_\Delta = (u_t - ww_x - v_x, v_t + uw_x + 3wu_x, w_t - u_x)$$

on $J^1(2,3)$. The corresponding equation is the dispersionless Boussinesq equation, see [44, 61].

Example 2.3 The universal hierarchy equation (UHE) is a subset:

$$\{u_{yy} - u_t u_{xy} + u_y u_{xt} = 0\} \subset J^2(3,1),$$

the coordinates in $J^2(3,1)$ being

$$x, y, t, u, u_x, u_y, u_t, u_{xx}, u_{xy}, u_{xt}, u_{yy}, u_{yt}, u_{tt},$$

given by the nonlinear differential operator

$$u \mapsto \frac{\partial^2 u}{\partial y^2} - \frac{\partial u}{\partial t} \cdot \frac{\partial^2 u}{\partial x \partial y} + \frac{\partial u}{\partial y} \cdot \frac{\partial^2 u}{\partial x \partial t},$$

see [93, 94].

2.1.1 $\mathscr{C}$-Differential Operators

Though all computations, by natural reasons, deal with finite objects, conceptually the central part of the theory is the infinite-dimensional space $J^\infty(n,m)$ whose coordinates are x^i, u^j, and all u^j_σ without limitations on the length of σ. We define smooth functions on $J^\infty(n,m)$ as smooth functions that depend on a finite but arbitrary number of jet variables. Let $\mathscr{F}(n,m;r)$ denote the space of smooth r-component vector functions. We shall also use the notation $\mathscr{F}(n,m) = \mathscr{F}(n,m;1)$.

The *total derivatives* are the operators:

$$D_i = \frac{\partial}{\partial x^i} + \sum_{j,\sigma} u^j_{\sigma i} \frac{\partial}{\partial u^j_\sigma} \tag{2.2}$$

that act from $\mathscr{F}(n,m)$ to $\mathscr{F}(n,m)$. Any linear differential operator $\Delta\colon \mathscr{F}(n,m;r) \to \mathscr{F}(n,m;r')$ that can be expressed in total derivatives is called a *$\mathscr{C}$-differential operator*. Thus, $\mathscr{C}$-differential operators of order k are matrices:

$$\Delta = \begin{pmatrix} \Delta^{11} & \dots & \Delta^{1r} \\ \dots\dots & \dots\dots\dots & \dots\dots \\ \Delta^{r'1} & \dots & \Delta^{r'r} \end{pmatrix}, \tag{2.3}$$

where

$$\Delta^{\alpha\beta} = \sum_{|\sigma| \le k} \Delta^{\alpha\beta}_\sigma D_\sigma, \qquad \Delta^{\alpha\beta}_\sigma \in \mathscr{F}(n,m),$$

and

$$D_\sigma = D_{i_1} \circ \dots \circ D_{i_s}, \qquad \sigma = i_1 \dots i_s.$$

If an equation is defined by a vector function $F = (F^1, \dots, F^r) \in \mathcal{F}(n, m; r)$, then its *infinite prolongation* $\mathcal{E} \subset J^\infty(n, m)$ is given by the infinite system:

$$D_\sigma(F^j) = 0, \qquad j = 1, \dots, r, \quad |\sigma| \geq 0.$$

Using Equality (2.2), let us describe infinite prolongations of the equations presented in Examples 2.1, 2.2, and 2.3.

Example 2.4 To obtain the first prolongation of the wave equation, we apply the total derivatives $D_0, D_1, \dots, D_n$ to (2.1) and obtain the system:

$$u_{000} = c\sum_{i=1}^{n} u_{0ii},$$

$$u_{001} = c\sum_{i=1}^{n} u_{1ii},$$

$$\dots$$

$$u_{00n} = c\sum_{i=1}^{n} u_{nii},$$

in $J^3(n+1, 1)$ that must be added to the initial equation $u_{00} = c\sum_{i=1}^{n} u_{ii}$. The infinite prolongation is defined by the infinite system:

$$u_{\sigma 00} = c\sum_{i=1}^{n} u_{\sigma ii}, \qquad |\sigma| \geq 0,$$

in $J^\infty(n+1, 1)$.

Example 2.5 To describe the infinite prolongation of the dispersionless Boussinesq equation, let us introduce the coordinates:

$$u_{i,j} = u_{\underbrace{t\dots t}_{i \text{ times}}\underbrace{x\dots x}_{j \text{ times}}}$$

in $J^\infty(2, 3)$ and $v_{i,j}$, $w_{i,j}$ defined in a similar way. Then the equations

$$u_{1,0} = w_{0,0}w_{0,1} + v_{0,1}, \quad v_{1,0} = -u_{0,0}w_{0,1} - 3w_{0,0}u_{0,1}, \quad w_{1,0} = u_{0,1}$$

together with

$$
\begin{aligned}
u_{2,0} &= w_{1,0}w_{0,1} + w_{0,0}w_{1,1} + v_{1,1}, & u_{1,1} &= w_{0,1}^2 + w_{0,0}w_{0,2} + v_{0,2},\\
v_{2,0} &= -u_{1,0}w_{0,1} + u_{0,0}w_{1,1} & &\\
&\quad - 3(w_{1,0}u_{0,1} + w_{0,0}u_{1,1}), & v_{1,1} &= -4u_{0,1}w_{0,1} - u_{0,0}w_{0,2} - 3w_{0,0}u_{0,2},\\
w_{2,0} &= u_{1,1}, & w_{1,1} &= u_{0,2}
\end{aligned}
$$

define the first prolongation in $J^2(2,3)$. The infinite prolongation is the system:

$$
\begin{aligned}
u_{i+1,j} &= \sum_{\alpha=0}^{i}\sum_{\beta=0}^{j}\binom{i}{\alpha}\binom{j}{\beta} w_{\alpha,\beta}w_{i-\alpha,j-\beta+1} + v_{i,j+1},\\
v_{i+1,j} &= \sum_{\alpha=0}^{i}\sum_{\beta=0}^{j}\binom{i}{\alpha}\binom{j}{\beta} (u_{\alpha,\beta}w_{i-\alpha,j-\beta+1} + 3w_{\alpha,\beta}u_{i-\alpha,j-\beta+1}),\\
w_{i+1,j} &= u_{i,j+1},
\end{aligned}
$$

where $i, j = 0, 1, \dots$

Example 2.6 Let us introduce the coordinates:

$$
u_{i,j,k} = u_{\underbrace{x\dots x}_{i\text{ times}}\underbrace{y\dots y}_{j\text{ times}}\underbrace{t\dots t}_{k\text{ times}}}
$$

in $J^\infty(3,1)$. Then the universal hierarchy equation takes the form:

$$
u_{0,2,0} = u_{0,0,1}u_{1,1,0} - u_{0,1,0}u_{1,0,1}
$$

in this coordinate system, while its infinite prolongation is given by the system

$$
\begin{aligned}
u_{i,j+2,k} = \sum_{\alpha=0}^{i}\sum_{\beta=0}^{j}\sum_{\gamma=0}^{k}\binom{i}{\alpha}\binom{j}{\beta}\binom{k}{\gamma} &(u_{\alpha,\beta,\gamma+1}u_{i-\alpha+1,j-\beta+1,k-\gamma}\\
&- u_{\alpha,\beta+1,\gamma}u_{i-\alpha+1,j-\beta,k-\gamma+1})
\end{aligned}
$$

where $i, j, k = 0, 1, \dots$

Local coordinates on the space of the infinite prolongation are called *internal coordinates* for the equation at hand. The space of smooth functions on the infinite prolongation will be denoted by $\mathscr{F}(\mathscr{E})$. The first part of the generality condition (see p. 6) is necessary for internal coordinates to exist: the differentials $d_\theta F^1, \dots, d_\theta F^r$ of the functions defining $\mathscr{E}$ must be linearly independent at any point $\theta \in \mathscr{E}$.

Example 2.7 One of possible choices of internal coordinates on the wave equation is

$$x^0, \dots, x^n, \qquad u^{(0)}_\sigma = u_\sigma, \qquad u^{(1)}_\sigma = u_{0\sigma},$$

where σ is an arbitrary multi-index containing the integers $1, \dots, n$ (but not 0).

Example 2.8 For the dispersionless Boussinesq equation, one can choose the internal coordinates x, t and

$$u_k = u_{\underbrace{x \dots x}_{k \text{ times}}}, \qquad v_k = v_{\underbrace{x \dots x}_{k \text{ times}}}, \qquad w_k = w_{\underbrace{x \dots x}_{k \text{ times}}}$$

where k is an arbitrary nonnegative integer.

Remark 2.2 The choice of internal coordinates described in Example 2.8 is typical for any system of *evolution equations*:

$$u^j_t = f^j\left(t, x^1, \dots, x^n, \dots, \frac{\partial^{|\sigma|} u^\alpha}{\partial x^\sigma}, \dots\right), \qquad j, \alpha = 1, \dots, m.$$

Namely, the "usual" choice is t, $x^1, \dots, x^n$ and the coordinates u^j_σ that correspond to $\partial^{|\sigma|} u^j / \partial x^\sigma$. But of course this choice is not unique (as any choice of local coordinates). For example, for the Burgers equation

$$u_t = u u_x + u_{xx}$$

we can choose the coordinates t, x, and u_k, as it was done for the Boussinesq equation, but also

$$u^{(0)}_k = u_{\underbrace{t \dots t}_{k \text{ times}}}, \qquad u^{(1)}_k = u_{x \underbrace{t \dots t}_{k \text{ times}}}$$

instead of u_k.

Example 2.9 An obvious choice of internal coordinates on UHE is x, y, t, and

$$u^{(0)}_{k,l} = u_{\underbrace{x \dots x}_{k \text{ times}} \underbrace{t \dots t}_{l \text{ times}}}, \qquad u^{(1)}_{k,l} = u_{y \underbrace{x \dots x}_{k \text{ times}} \underbrace{t \dots t}_{l \text{ times}}},$$

but one can also choose

$$u^{(0)}_l = u_{\underbrace{t \dots t}_{l \text{ times}}}, \qquad u^{(x)}_{k,l} = u_{\underbrace{x \dots x}_{k \text{ times}} \underbrace{t \dots t}_{l \text{ times}}}, \qquad u^{(y)}_{k,l} = u_{\underbrace{y \dots y}_{k \text{ times}} \underbrace{t \dots t}_{l \text{ times}}},$$

where $l \geq 0$, $k \geq 1$.

For a proper choice of internal coordinates, the equation at hand must be presented in the *passive orthonomic form*, see [5, 97]. This important concept originates from the works by M. Janet and Ch. Riquier [52, 123] (see also [116]) and means the following. Consider the space $J^\infty(n; m)$ and denote by

$$\mathrm{AD} = \{\, u^j_\sigma \mid |\sigma| \geq 0,\ j = 1, \dots, m\}$$

the set of adapted coordinates in this space. A ranking of this set is a linear ordering $\preceq$ such that:

- $p \prec D_i(p)$ for any $p \in \mathrm{AD}$ and any total derivative D_i;
- the relation $p \prec q$ implies $D_i(p) \prec D_i(q)$.

Choose a ranking and assume that the equation under consideration is defined by the system:

$$F^\alpha(\dots, x^i, \dots, u^j_\sigma, \dots) = 0, \qquad 1 \leq \alpha \leq l. \tag{2.4}$$

Let $\Sigma_F \subset \mathrm{AD}$ be the set of variables u^j_σ on which the functions F^α depend; then a variable u^k_μ is called a *leading derivative* if it is a maximal element in Σ_F with respect to the chosen ranking. Resolve Eqs. (2.4) with respect to the leading derivatives:

$$u^j_\mu = \Phi^j_\mu, \tag{2.5}$$

where Φ^j_μ are some other functions on the ambient jet space, and consider the infinite prolongation $\mathcal{E}$ presented in the form

$$u_{\mu\sigma} = D_\sigma(\Phi^j_\mu). \tag{2.6}$$

Then the variables at the left-hand side are called *principal derivatives* all the others being *parametric* ones.

Consider System (2.6) and say that it is:

- *triangular* if for every leading derivative in this system there exists exactly one equation, where this derivative appears in the left-hand side;
- *autoreduced* if no principal derivative appears in the right-hand side;
- *orthonomic* if it is both triangular and autoreduced.

Example 2.10 In all the above examples, the equations under consideration were presented in the orthonomic passive form. But if we rewrite the UHE as

$$u_t = \frac{u_{yy} + u_y u_{xt}}{u_{xy}}$$

this will not be the case.

When System (2.5) consists of more than one equation, it may contain hidden dependencies between parametric variables (that lead to compatibility conditions). If such dependencies do exist, the system is called *active*; otherwise, it is called *passive*.

Example 2.11 All the systems above were passive. But, for example,

$$u_x = f(x, y, u), \qquad u_y = g(x, y, u)$$

is active.

Any $\mathscr{C}$-differential operator Δ can be restricted to the space of infinite prolongation. Efficiently, this means that Δ can be rewritten in terms of internal coordinates. The restriction is denoted by $\Delta|_{\mathscr{E}}$. The spaces of functions on $\mathscr{E}$ with the values in $\mathbb{R}^l$, i.e., $\mathbb{R}^l$-valued functions expressed in terms of internal coordinates, will be denoted by $\mathscr{F}(\mathscr{E}; l)$. In local coordinates, to restrict a $\mathscr{C}$-differential operator to $\mathscr{E}$ means to rewrite it in terms of parametric derivatives. Since $\mathscr{C}$-differential operators are expressed in terms of total derivatives, it is essential to describe the fields $D_i|_{\mathscr{E}}$. Consider examples.[1]

Example 2.12 Let internal coordinates for the linear wave equation be chosen as in Example 2.7. Then

$$D_0 = \frac{\partial}{\partial x^0} + \sum_{|\sigma|\geq 0} u_\sigma^{(1)} \frac{\partial}{\partial u_\sigma^{(0)}} + c \sum_{|\sigma|\geq 0} \left(\sum_{i=1}^{n} u_{ii\sigma}^{(0)} \right) \frac{\partial}{\partial u_\sigma^{(1)}}$$

and

$$D_i = \frac{\partial}{\partial x^i} + \sum_{|\sigma|\geq 0} u_\sigma^{(0)} \frac{\partial}{\partial u_\sigma^{(0)}} + \sum_{|\sigma|\geq 0} u_{\sigma i}^{(0)} \frac{\partial}{\partial u_\sigma^{(1)}}$$

for $i = 1, \dots, n$. Recall that σ contains the integers $1, \dots, n$ in this example.

Example 2.13 Consider the dispersionless Boussinesq equation and choose internal coordinates as in Example 2.8. Then

$$D_x = \frac{\partial}{\partial x} + \sum_{k\geq 0} \left(u_{k+1} \frac{\partial}{\partial u_k} + v_{k+1} \frac{\partial}{\partial v_k} + w_{k+1} \frac{\partial}{\partial w_k} \right)$$

and

$$D_t = \frac{\partial}{\partial t} + \sum_{k\geq 0} \left(D_x^k(w_0 w_1 + v_1) \frac{\partial}{\partial u_k} - D_x^k(u_0 w_1 + 3 w_0 u_1) \frac{\partial}{\partial v_k} + u_{k+1} \frac{\partial}{\partial w_k} \right).$$

[1] Usually, we preserve the notation D_i for the restriction $D_i|_{\mathscr{E}}$.

Remark 2.3 For a general evolution equation with the "standard" choice of internal coordinates (see Remark 2.2), the total derivatives acquire the form:

$$D_i = \frac{\partial}{\partial x^i} + \sum_{\sigma,j} u^j_{\sigma i} \frac{\partial}{\partial u^j_\sigma}$$

and

$$D_t = \frac{\partial}{\partial t} + \sum_{\sigma,j} D_\sigma(f^j) \frac{\partial}{\partial u^j_\sigma}.$$

But if the choice is different (like for the Burgers equation in the same remark), then we have

$$u_{xx} = -u_t - uu_x$$

and

$$D_t = \frac{\partial}{\partial t} + \sum_{k\geq 0} \left(u^{(0)}_{k+1} \frac{\partial}{\partial u^{(0)}_k} + u^{(1)}_{k+1} \frac{\partial}{\partial u^{(1)}_k} \right),$$

$$D_x = \frac{\partial}{\partial x} + \sum_{k\geq 0} u^{(1)}_k \frac{\partial}{\partial u^{(0)}_k} + \sum_{k\geq 0} D^k_t \left(-u^{(0)}_1 - u^{(0)}_0 u^{(1)}_0 \right) \frac{\partial}{\partial u^{(1)}_k}.$$

Example 2.14 In Example 2.9, we considered two different choices of internal coordinates on UHE. The first one corresponds to the representation of the equation in the form:

$$u_{yy} = u_t u_{xy} - u_y u_{xt}.$$

The total derivatives for this choice are

$$D_x = \frac{\partial}{\partial x} + \sum_{k,l} \left(u^{(0)}_{k+1,l} \frac{\partial}{\partial u^{(0)}_{k,l}} + u^{(1)}_{k+1,l} \frac{\partial}{\partial u^{(1)}_{k,l}} \right),$$

$$D_t = \frac{\partial}{\partial t} + \sum_{k,l} \left(u^{(0)}_{k,l+1} \frac{\partial}{\partial u^{(0)}_{k,l}} + u^{(1)}_{k,l+1} \frac{\partial}{\partial u^{(1)}_{k,l}} \right),$$

$$D_y = \frac{\partial}{\partial y} + \sum_{k,l} u^{(1)}_{k,l} \frac{\partial}{\partial u^{(0)}_{k,l}} + \sum_{k,l} D^k_x D^l_t \left(u^{(0)}_{0,1} u^{(1)}_{1,0} - u^{(1)}_{0,0} u^{(0)}_{1,1} \right) \frac{\partial}{\partial u^{(1)}_{k,l}}.$$

The other choice corresponds to the representation:

$$u_{xy} = \frac{u_{yy} + u_y u_{xt}}{u_t}$$

and

$$D_t = \frac{\partial}{\partial t} + \sum_l u^{(0)}_{l+1} \frac{\partial}{\partial u^{(0)}_l} + \sum_{k,l} \left(u^{(x)}_{k,l+1} \frac{\partial}{\partial u^{(x)}_{k,l}} + u^{(y)}_{k,l+1} \frac{\partial}{\partial u^{(y)}_{k,l}} \right),$$

$$D_x = \frac{\partial}{\partial x} + \sum_l u^{(x)}_{1,l} \frac{\partial}{\partial u^{(0)}_l} + \sum_{k,l} u^{(x)}_{k+1,l} \frac{\partial}{\partial u^{(x)}_{k,l}} + \sum_{k,l} D_y^{k-1} D_t^l \left(\frac{u^{(y)}_{2,0} + u^{(y)}_{1,0} u^{(x)}_{1,1}}{u^{(0)}_1} \right) \frac{\partial}{\partial u^{(y)}_{k,l}},$$

$$D_y = \frac{\partial}{\partial y} + \sum_l u^{(y)}_{1,l} \frac{\partial}{\partial u^{(0)}_l} + \sum_{k,l} D_x^{k-1} D_t^l \left(\frac{u^{(y)}_{2,0} + u^{(y)}_{1,0} u^{(x)}_{1,1}}{u^{(0)}_1} \right) \frac{\partial}{\partial u^{(x)}_{k,l}} + \sum_{k,l} u^{(y)}_{k+1,l} \frac{\partial}{\partial u^{(y)}_{k,l}}.$$

in this case.

2.1.2 *The Linearization Operator and Its Adjoint*

Let $F \in \mathscr{F}(n,m;r)$ be a vector function. Its *linearization* is the $\mathscr{C}$-differential operator $\ell_F \colon \mathscr{F}(n,m;m) \to \mathscr{F}(n,m;r)$ given by the formula:

$$\ell_F = \begin{pmatrix} \sum_\sigma \partial F^1/\partial u^1_\sigma D_\sigma & \dots & \sum_\sigma \partial F^1/\partial u^m_\sigma D_\sigma \\ \dots & \dots & \dots \\ \sum_\sigma \partial F^r/\partial u^1_\sigma D_\sigma & \dots & \sum_\sigma \partial F^r/\partial u^m_\sigma D_\sigma \end{pmatrix}.$$

If $\mathscr{E}$ is the equation defined by F, we use the notation:

$$\ell_{\mathscr{E}} = \ell_F|_{\mathscr{E}}.$$

Let $\Delta \colon \mathscr{F}(n,m;r) \to \mathscr{F}(n,m;r')$ be a $\mathscr{C}$-differential operator of the form (2.3). Its *adjoint* is the operator $\Delta^* \colon \mathscr{F}(n,m;r') \to \mathscr{F}(n,m;r)$ given by

$$\Delta^* = \begin{pmatrix} (\Delta^{11})^* & \dots & (\Delta^{r'1})^* \\ \dots & \dots & \dots \\ (\Delta^{1r})^* & \dots & (\Delta^{r'r})^* \end{pmatrix},$$

where

$$(\Delta^{\alpha\beta})^* = \left(\sum_\sigma \Delta^{\alpha\beta}_\sigma D_\sigma \right)^* = \sum_\sigma (-1)^{|\sigma|} D_\sigma \circ \Delta^{\alpha\beta}_\sigma.$$

In particular, the adjoint to ℓ_F is given by

$$\ell_F^* = \begin{pmatrix} \sum_\sigma (-1)^{|\sigma|} D_\sigma \circ \partial F^1/\partial u_\sigma^1 & \dots & \sum_\sigma (-1)^{|\sigma|} D_\sigma \circ \partial F^r/\partial u_\sigma^1 \\ \dots\dots & \dots & \dots\dots \\ \sum_\sigma (-1)^{|\sigma|} D_\sigma \circ \partial F^1/\partial u_\sigma^m & \dots & \sum_\sigma (-1)^{|\sigma|} D_\sigma \circ \partial F^r/\partial u_\sigma^m \end{pmatrix}.$$

A $\mathscr{C}$-differential operator Δ and its adjoint Δ^* are related to each other by the *Green formula*:

$$\langle \Delta(\varphi), \psi \rangle - \langle \varphi, \Delta^*(\psi) \rangle = \sum_{i=1}^{n} D_i(a_i), \tag{2.7}$$

where $\varphi \in \mathscr{F}(n, m; r)$, $\psi \in \mathscr{F}(n, m; r')$, $\langle v, v' \rangle = \sum_i v_i v_i'$, and $a_1, \dots, a_n \in \mathscr{F}(n, m)$.

Remark 2.4 As it follows from Chap. 3, the right-hand side of the Green formula is actually of the form $d_h \omega_\Delta(\varphi, \psi)$, where $\omega_\Delta(\varphi, \psi)$ is an $(n-1)$-horizontal differential form depending on Δ, φ, and ψ. Moreover, it can be shown that the correspondence $(\varphi, \psi) \mapsto \omega_\Delta(\varphi, \psi)$ is a $\mathscr{C}$-differential operator both in φ and in ψ.

Consider examples.

Example 2.15 The linear wave equation is defined by the function (2.1):

$$F = u_{00} - c \sum_{i=1}^{n} u_{ii}.$$

The linearization of F is

$$\ell_F = D_t^2 - c \sum_{i=1}^{n} D_i^2,$$

where the total derivatives are considered on the ambient jet space.

Remark 2.5 The differential operator Δ_F corresponding to F is a linear one. Linearizations of linear differential operators coincide with their lifts (see Example 1.1).

The adjoint operator is of the form:

$$\ell_F^* = D_t^2 - \sum_{i=1}^{n} D_i^2 \cdot c = D_t^2 - c \sum_{i=1}^{n} D_i^2$$

and thus coincides with ℓ_F. Consequently, due to Remark 1.4, the linear wave equation is an Euler-Lagrange one with the Lagrangian density:

$$L = \left(-u_0^2 + c \sum_{i=1}^{n} u_i^2 \right) dt \wedge dx^1 \wedge \dots \wedge dx^n.$$

To obtain the restrictions $\ell_{\mathcal{E}} = \ell_F|_{\mathcal{E}}$ and $\ell_{\mathcal{E}}^* = \ell_F^*|_{\mathcal{E}}$, it suffices to understand the total derivatives in the sense of Example 2.12.

Example 2.16 The vector function $F = (F^1, F^2, F^3)$, where

$$F^1 = u_t - ww_x - v_x, \quad F^2 = v_t + uw_x + 3wu_x, \quad F^3 = w_t - u_x$$

defines the dispersionless Boussinesq equation (see Example 2.5). Then

$$\ell_F = \begin{pmatrix} D_t & -D_x & -w_x - wD_x \\ w_x + 3wD_x & D_t & 3u_x + uD_x \\ -D_x & 0 & D_t \end{pmatrix},$$

while

$$\ell_F^* = \begin{pmatrix} -D_t & w_x - 3D_x \circ w & D_x \\ D_x & -D_t & 0 \\ -w_x + D_x \circ w & 3u_x - D_x \circ u & -D_t \end{pmatrix} = \begin{pmatrix} -D_t & -2w_x - 3wD_x & D_x \\ D_x & -D_t & 0 \\ wD_x & 2u_x - uD_x & -D_t \end{pmatrix}.$$

Again, to obtain the operators $\ell_{\mathcal{E}}$ and $\ell_{\mathcal{E}}^*$, one needs to substitute the total derivatives computed in Example 2.13.

Example 2.17 The Burgers equation is defined by the function $F = u_t - uu_x - u_{xx}$. Consequently,

$$\ell_{\mathcal{E}} = D_t - u_x - uD_x - D_x^2$$

and

$$\ell_{\mathcal{E}}^* = -D_t - u_x + D_x \circ u - D_x^2 = -D_t + uD_x - D_x^2,$$

where the total derivatives are chosen in the "standard" way:

$$D_x = \frac{\partial}{\partial x} + \sum_{k \geq 0} u_{k+1} \frac{\partial}{\partial u_k},$$

$$D_t = \frac{\partial}{\partial t} + \sum_{k \geq 0} D_x^k (u_0 u_1 + u_2) \frac{\partial}{\partial u_k}.$$

(cf. Remark 2.3).

Example 2.18 The UHE is given by the function:

$$F = u_{yy} - u_t u_{xy} + u_y u_{xt}$$

(see Example 2.3). Then

$$\ell_{\mathcal{E}} = D_y^2 - u_t D_x D_y + u_y D_x D_t - u_{xy} D_t + u_{xt} D_y$$

and

$$\ell_{\mathcal{E}}^* = D_y^2 - D_x D_y \circ u_t + D_x D_t \circ u_y + D_t \circ u_{xy} - D_y \circ u_{xt}$$
$$= D_y^2 - u_t D_x D_y + u_y D_x D_t - 2u_{xt} D_y + 2u_{xy} D_t.$$

2.2 CDE Implementation

Let us now demonstrate how the constructions of the previous section are realized by means of the CDE software.

For the reader's convenience, all CDE commands (or functions) are written in a section on page 243. The list of example programs is also available on page 247.

2.2.1 CDE Jet Space

CDE is first of all a program which is able to create a *finite-order jet space* inside Reduce. To this aim, issue the command:

```
load_package cde;
```

Then, CDE needs to know the variables and the maximal order of derivatives. The input can be organized as in the following example with two independent variables and two dependent variables:

```
indep_var:={x,t};
dep_var:={u,v};
total_order:=10;
```

Here:

- `indep_var` is the list of independent variables;
- `dep_var` is the list of dependent variables;
- `total_order` is the maximal order of derivatives.

The above newly created variables are *global variables*.[2] The list of all global variables in the system can be found on page 251.

[2] More precisely, they are fluid variables. See Reduce manual [45] for more details.

We recall that there are two modes of interpreting commands in Reduce: the `algebraic` mode and the `symbolic` mode (see Sect. A.2 in the Appendix). CDE is programmed in symbolic mode, for greater efficiency and flexibility, but it is used in algebraic mode: this means that variables shall be copied from algebraic to symbolic mode. In CDE, global symbolic mode variables always have the characters `'!*'` in the end of their name. The variables which correspond to `indep_var` and `dep_var` in symbolic mode are `indep_var!*` and `dep_var!*`. Integer values are accessible in both modes.

Two more parameters can be set for convenience:

```
statename:="jetuv_state.red";
resname:="jetuv_res.red";
```

These are the name of the output file for recording the internal state of CDE (e.g., for debugging purposes) and the name of the file containing results of the computation.

The main routine in `cde.red` is called as follows:

```
cde({indep_var,dep_var,{},total_order},{});
```

Here the two empty lists are placeholders; they are important for computations with odd variables/differential equations. The function `cde` defines derivative symbols of the type:

```
u_x,v_t,u_2xt,v_xt,v_2x3t,...
```

Note that the symbol `v_tx` does not exist in the CDE jet space. Indeed, introducing all possible permutations of independent variables in indices would increase the complexity of every operation and eventually slow down computations.

A useful list is generated by CDE, namely, `all_der_id`, the list of all identifiers of even variables; the number of independent variables is stored in `n_indep_var`. Other lists are generated by CDE,[3] but they are accessible in Reduce symbolic mode only. See the list of global variables on page 251.

It can also be useful to inspect the output generated by the function `cde` and the above lists in particular. All that data can be saved by the function:

```
save_cde_state(statename);
```

CDE has a few procedures involving the jet space, namely:

- `jet_fiber_dim(jorder)` returns the number of derivative coordinates u^i_σ with $|\sigma|$ equal to `jorder`;
- `jet_dim(jorder)` returns the number of derivative coordinates u^i_σ with $0 \leq |\sigma| \leq r$ and r equal to `jorder`;
- `selectvars(par,orderofder,depvars,vars)` returns all derivative coordinates (even, if `par=0`, odd, if `par=1`, see below for odd coordinates)

[3]They are `all_mind_table!*`, `all_der_mind!*`, `all_mind!*`, `i2m_jetspace!*`, `i2o_jetspace!*`, `m2o_jetspace!*`

of order `orderofder` of the list of the dependent variables `depvars` which belong to the set of derivative coordinates `vars`.

The function `cde` defines total derivatives truncated at the order `total_order`; this means that the sum in (2.2) is extended to all σ such that $|\sigma|$ is less than or equal to `total_order`. Total derivatives are introduced through a `CDIFF` function that defines super vector fields, using data prepared by `CDE`.

The total derivative of an argument `phi` is invoked as follows:

```
td(phi,x,2);
td(phi,x,t,3);
```

the syntax closely follows Reduce syntax for standard derivatives `df`; the above expressions translate to $D_x D_x \varphi$, or $D_{\{2,0\}}\varphi$ in multi-index notation, and $D_x D_t^3 \varphi$, or $D_{\{1,3\}}\varphi$, respectively. Internally, total derivatives are denoted by functions whose names are built up by the string `id_tot_der!*` and the identifier of each independent variable, and they are stored in the list `tot_der!*`.

When in total derivatives there is a coefficient of order higher than maximal (i.e., for $|\sigma|$ =`total_order`), this is replaced by the identifier `letop`,[4] which is a function that depends on independent variables. If such a function (or its derivatives) appears during computations, this means that the computation requires a jet space with a higher order. All results of computations with total derivatives are scanned for the presence of `letop`; if the `letop` identifier is detected, the computation is stopped with an error message. This implies that we need to extend the order of the jet space, just by increasing the number `total_order`. This can be done automatically: more precisely, any `CDE` program can be run through a script that detects error messages containing the keyword `letop`. In this case, the script can rerun the program with a new value of `total_order` one unit higher than the previous one. See, for example, the Linux script `rrr.sh` (included in `CDE`).

The function `check_letop` checks an expression for the presence of the identifier `letop` or its derivatives. If you wish to switch off this kind of check in order to increase the speed, the switch `checkord` must be set off:

```
off checkord;
```

The computation of total derivatives of a huge expression can be extremely time and resource consuming. In some cases it is a good idea to disable the expansion of the total derivative and leave an expression of the type $D_\sigma \varphi$ as indicated. This is achieved by the command:

```
noexpand_td();
```

If you wish to restore the default behavior, do

```
expand_td();
```

[4]In Dutch "let op" means "pay attention."

CDE can also compute on jets of supermanifolds (see [51, 57, 82]). In this book we will only deal with odd variables as far as they are needed to compute integrability operators.

The input can be organized as follows:

```
indep_var:={x,t};
dep_var:={u,v};
odd_var:={p,q};
total_order:=10;
```

Here odd_var is the list of odd variables. The call

```
cde({indep_var,dep_var,odd_var,total_order},{});
```

will create the jet space of the supermanifold described by the independent variables and the even and odd dependent variables, up to the order total_order. This includes the list all_odd_id and the number of all odd variables and their derivatives n_all_ext.[5]

Total derivatives truncated at the order total_order will also include odd derivatives; the considerations on expansion and letop apply in this case too.

Odd variables can appear in anticommuting products; this is represented as

```
ext(p,p_2xt),ext(p_x,q_t,q_x2t),...
```

Internally, odd variables and their derivatives are indexed by integers, and the odd products of odd variables are represented, for example, as

```
ext(2),ext(3,23),ext(1,3,5),...
```

The above representation is not intended for normal users. Note that p and ext(p) are just the same. The odd product of two expressions phi and psi is achieved by the CDIFF function:

```
odd_product(phi,psi);
```

Note that it does not hold in the system that ext(p_2xt,p) is automatically expanded to -ext(p,p_2xt) for performance reasons. Reduce always represents an odd product in a unique way which depends on the internal representation of odd variables. The user should always use the function odd_product in order to write the anticommutative product of two or more odd variables. In the opposite case, there is the risk of introducing a new expression which is not related to what is in the system.

The derivative of an expression phi with respect to an odd variable p is achieved by

```
df_odd(phi,p);
```

[5]There are exclusively symbolic mode lists as well: all_odd_mind!*, all_odd_id!*, i2m_jetspace_odd!*, i2o_jetspace_odd!*

2.2.2 CDE and Differential Equations

In this section, we will solve Problem 1.3.

CDE needs the equation in a passive orthonomic form. Indeed, CDE is not able to compute if there are hidden dependencies (or integrability conditions) in the differential consequences of the equation at hand (see the discussion on page 34). See the software package [15] for the problem of transforming an equation to a passive orthonomic form.

When the equation is in passive orthonomic form, one or more derivatives are equated to the right-hand side expressions. Recall (see Sects. 2.1 and 2.1.1, in particular) that the left-hand side derivatives are called *principal*, and the remaining derivatives are called *parametric*. Parametric coordinates are internal coordinates on the equation manifold and its differential consequences $\mathscr{E}$. Here $\mathscr{E}$ is regarded as the graph of a vector function in the ambient jet space. Principal derivatives are determined by the differential equation and its differential consequences. For scalar evolution equations with two independent variables $u_t = f(u, u_x, u_{2x}, \ldots)$, parametric derivatives are usually of the type $(u, u_x, u_{xx}, \ldots)$. The input is formed as follows[6]:

```
% left-hand side of the differential equation
  principal_der:={u_t};
% right-hand side of the differential equation
  de:={f(u,u_x,u_2x,...)};
```

Systems of PDEs are input in the same way: of course, the above two lists must have the same length.

The main routine in CDE is called as follows:

```
cde({indep_var,dep_var,{},total_order},
  {principal_der,de,{},{}});
```

Here the three empty lists are placeholders; they are important for computations with odd variables. The function `cde` computes principal and parametric derivatives of even and odd variables, they are stored in the lists `all_parametric_der`, `all_principal_der`.

The function `cde` also restricts total derivatives truncated at the order whose value is equal to `total_order` to the (even and odd) equation; this means that total derivatives in CDE are always tangent to the equation manifold. Their coordinate expressions are of the form:

$$D_\lambda = \frac{\partial}{\partial x^\lambda} + \sum_{u^i_\sigma \text{ parametric}} u^i_{\sigma\lambda} \frac{\partial}{\partial u^i_\sigma} + \sum_{p^i_\sigma \text{ parametric}} p^i_{\sigma\lambda} \frac{\partial}{\partial p^i_\sigma}, \tag{2.8}$$

[6] Lines beginning with % are comments in Reduce.

where x^λ are independent variables, u^i are dependent variables, p^j are odd variables, and σ is a multi-index. It can happen that $u^i_{\sigma\lambda}$ (or $p^j_{\sigma\lambda}$) is principal and must be replaced with differential consequences of the equation. Such differential consequences are called *primary differential consequences*, and are computed; in general they will depend on other, possibly new, principal derivatives. They must be replaced by the corresponding differential consequences. Such newly appearing differential consequences are called *secondary differential consequences*. Secondary differential consequences may give rise to tertiary differential consequences, and so on. If the equation is in passive orthonomic form, the system of all differential consequences (up to the maximal order `total_order`) must be solvable in terms of parametric derivatives only. The function `cde` automatically computes all necessary and sufficient differential consequences which are needed to solve the system. In the process of computing and solving the system of all differential consequences, several lists are created in symbolic mode; the reader is invited to consult the list of all global variables on page 251.

The solved system is available in the form of Reduce replacement rules in the variable `repprincparam_der`.

The syntax and properties (expansion and `letop`) of total derivatives remain the same. In the above example,

```
td(u,t);
```

returns

```
f(u,u_x,u_2x,...).
```

It is possible to deal with mixed systems of even and odd variables. For example, we can add an odd scalar equation by adding a new odd variable `q` to our previous equation as follows:

```
odd_var:={q};
principal_odd:={q_t};
de_odd:={g(u,u_x,u_2x,...,q,q_x,q_2x,...)};
```

The function g must be polynomial in odd variables, as it must be a *superfunction* on a supermanifold. Of course, products between odd variables must be input using the `odd_product` command. The main routine in `CDE` is in this case called as follows:

```
cde({indep_var,dep_var,odd_var,total_order},
{principal_der,de,principal_odd,de_odd});
```

As above, the list of principal and parametric odd variables are created and stored in the variables `all_parametric_odd, all_principal_odd`.

The same mechanism of restricting total derivatives to the differential equation holds for mixed even and odd equations. The odd part of the solved system of differential consequences is available in the form of Reduce replacement rules in `repprincparam_odd`.

2.2.3 CDE and $\mathscr{C}$-Differential Operators

In this section, we will solve Problems 1.1, 1.2, and 1.4.

Consider a vector-valued $\mathscr{C}$-differential operator $\Delta = (\ldots, \Delta^j, \ldots)$:

$$\Delta^j(\varphi_1, \ldots, \varphi_h) = \sum_{\substack{\sigma_1,\ldots,\sigma_h,\\ i_1,\ldots,i_h}} a^{\sigma_1,\ldots,\sigma_h,\,j}_{i_1\cdots i_h} D_{\sigma_1}\varphi_1^{i_1} \cdots D_{\sigma_h}\varphi_h^{i_h}. \tag{2.9}$$

The above $\mathscr{C}$-differential operator in CDE must be declared as follows:

```
mk_cdiffop(opname,num_arg,length_arg,length_target);
```

where:

- `opname` is the name of the operator (Δ in (2.9));
- `num_arg` is the number of arguments (h in (2.9));
- `length_arg` is the list of lengths of the arguments, e.g., in (2.9) one needs a list of h items `{k_1,...,k_h}`, each corresponding to number of components of the vector functions $\varphi_j^{i_j}$ to which the operator is applied;
- `length_target` is the number of components of the image vector function (the range of the index j in (2.9)).

The above parameters of the operator `opname` are stored in the *property list*[7] of the identifier `opname`. This means that if one would like to know how many arguments has the operator `opname`, the answer will be the output of the command:

```
get('cdnarg,cdiff_op);
```

and the same for the other parameters. The syntax for one component of the operator `opname` is

```
opname(j,i1,...,ih,phi1,...,phih);
```

A $\mathscr{C}$-differential operator of the type of (2.9) is *skew-symmetric* if and only if its coefficients $a^{\sigma_1,\ldots,\sigma_h,\,j}_{i_1\cdots i_h}$ are skew-symmetric with respect to the exchange of pairs of indices $^{\sigma_p}_{i_p}$ and $^{\sigma_q}_{i_q}$. Skew-symmetric operators (2.9) are represented as homogeneous vector-valued superfunctions on a supermanifold, where odd coordinates q^i_σ correspond to total derivatives $D_\sigma\varphi^i$ (see [40, 51, 57]). Indeed, from the viewpoint of computer algebra systems, it is much easier to work with polynomials than with operators. The isomorphism between the spaces of operators and of superfunctions is given by

$$\left(\sum_{\substack{\sigma_1,\ldots,\sigma_h,\\ i_1,\ldots,i_h}} a^{\sigma_1,\ldots,\sigma_h,\,j}_{i_1\cdots i_h} D_{\sigma_1}\varphi_1^{i_1} \cdots D_{\sigma_h}\varphi_h^{i_h}\right) \longrightarrow \left(\sum_{\substack{\sigma_1,\ldots,\sigma_h,\\ i_1,\ldots,i_h}} a^{\sigma_1,\ldots,\sigma_h,\,j}_{i_1\cdots i_h} q^{i_1}_{\sigma_1} \cdots q^{i_h}_{\sigma_h}\right) \tag{2.10}$$

[7]The property list is a Lisp concept, see [106] for details.

where q^i_σ is the derivative of an odd dependent variable (and an odd variable itself).

A vector-valued superfunction in CDE must be declared as follows:

```
mk_superfun(sfname,deg,length_target);
```

where:

- `sfname` is the name of the superfunction;
- `deg` is the degree of the superfunction, e.g., h in (2.10);
- `length_target` is the number of components of the image vector (the range of the index j in (2.10)).

The syntax for one component of the superfunction `sfname` is

```
sfname(j);
```

CDE is able to deal with $\mathscr{C}$-differential operators in both formalisms and provides conversion utilities between superfunctions of degree 1 and $\mathscr{C}$-differential operators with one argument:

- `conv_cdiff2superfun(cdop,superfun)`
- `conv_superfun2cdiff(superfun,cdop)`

In the first case, a $\mathscr{C}$-differential operator `cdop` is converted into a vector-superfunction `superfun` with the same properties, and conversely.

The linearization of a purely even vector function and the adjoint of a $\mathscr{C}$-differential operator with one argument are computed by means of the correspondence (2.10). Indeed, for symbolic software it is much easier to deal with an algebraic expression in odd variables rather than operators. Operations between general $\mathscr{C}$-differential operators are currently being programmed in CDE and will be available soon.

A vector function must be introduced in CDE as a list of scalar functions:

```
fun:={fun1,fun2,...};
```

Then its linearization is achieved by

```
ell_function(fun,lfun);
```

where `lfun` is automatically declared as a $\mathscr{C}$-differential operator with the appropriate parameters. Moreover, the above command creates a superfunction `lfun_sf` that corresponds to the $\mathscr{C}$-differential operator `lfun`.

The command

```
adjoint_cdiffop(lfun,lfun_star);
```

computes the adjoint `lfun_star` of `lfun` and introduces a superfunction whose identifier has the suffix `_sf`: `lfun_star_sf`.

Once the linearization and its adjoint are computed, in order to do computations with symmetries and conservation laws (see below), such operators must be restricted to the corresponding equation. This can be achieved with the following steps:

(1) compute the linearization and its adjoint of a PDE of the form $F = 0$ *in the free jet space* (i.e., no equation), and save them in the form of vector-valued superfunctions;
(2) start a new computation with the given *even* PDE as a constraint on the (even) jet space;
(3) load the superfunctions of item 1;
(4) restrict them to the even PDE.

Only the last step needs to be explained. If we are considering, e.g., the Boussinesq equation, then u_t and its differential consequences (i.e., the principal derivatives) are not automatically expanded to the right-hand side of the equation and its differential consequences. At the moment this step is not fully automatic. More precisely, only principal derivatives which appear as coefficients in total derivatives can be replaced by their expression. The lists of such derivatives with the corresponding expressions are `repprincparam_der` and `repprincparam_odd`. They are in the format of Reduce replacement list and can be used in `let rules`. If the linearization or its adjoint happen to depend on another principal derivative, this must be computed separately. A forthcoming release of Reduce will automatize this procedure.

However, note that for evolution equations, this step is trivial, as the restriction of linearization and its adjoint on the given PDE will only affect total derivatives, which are restricted by `CDE` to the PDE.

2.3 Examples

Let us illustrate the above constructions with typical examples. We will focus on computing linearization and its adjoint.

2.3.1 Korteweg-de Vries Equation

We present the KdV equation in the form:

$$u_t = uu_x + u_{xxx}$$

and choose the functions:

$$x, t, u, \dots, u_k = u_{\underbrace{x\dots x}_{k \text{ times}}}, \dots \tag{2.11}$$

for internal coordinates. The total derivatives on $\mathscr{E}$ are

$$D_x = \frac{\partial}{\partial x} + \sum_k u_{k+1}\frac{\partial}{\partial u_k},$$

$$D_t = \frac{\partial}{\partial t} + \sum_k D_x^k(uu_1 + u_3)\frac{\partial}{\partial u_k}.$$

The linearization is of the form:

$$\ell_{\mathscr{E}} = D_t - u_1 - uD_x - D_x^3,$$

while its adjoint is given by

$$\ell^*_{\mathscr{E}} = -D_t + uD_x + D_x^3.$$

The CDE implementation can be found in the program file kdv_ell1.red. In the free jet space

```
indep_var:={x,t};
dep_var:={u};
odd_var:={p};
total_order:=10;
```

Let us introduce the vector function whose zero set is the KdV equation:

```
f_kdv:={u_t - u*u_x - u_3x};
```

The following command assigns the linearization operator ℓ_{KdV} of the vector function `f_kdv` to the identifier `lkdv`

```
ell_function(f_kdv,lkdv);
```

moreover, a superfunction `lkdv_sf` is also defined as the vector superfunction corresponding to ℓ_{KdV}. Indeed, the following command[8]:

```
lkdv_sf(1);

q_t - q*u_x - q_x*u  - q_3x;
```

shows the vector superfunction corresponding to ℓ_{KdV}. To compute the value of the $(1, 1)$ component of the matrix ℓ_{KdV} applied to an argument `psi` do

```
lkdv(1,1,psi);
```

In order to check that the result is correct, one could define the linearization as a $\mathscr{C}$-differential operator and then check that the corresponding superfunctions are the same:

[8]Here and in what follows a command is separated by its output by means of a blank line in between them.

```
mk_cdiffop(lkdv2,1,{1},1);
for all phi let lkdv2(1,1,phi)=
   td(phi,t) - u_x*phi - u*td(phi,x) - td(phi,x,3);

conv_cdiff2superfun(lkdv2,lkdv2_sf);
lkdv2_sf(1) - lkdv_sf(1);
```

the result of the two last commands must be zero.

The adjoint of `lkdv` can be computed and assigned to the identifier `lkdv_star` by the command:

```
adjoint_cdiffop(lkdv,lkdv_star);
```

Again, the associated vector superfunction `lkdv_star_sf` is computed, with values:

```
lkdv_star_sf(1);

 - q_t + u*q_x  + q_3x
```

Again, the above operator can be checked for correctness.

2.3.2 Dispersionless Boussinesq System

Recall that the Boussinesq equation can be presented as a system in $J^1(2,3)$ (see Examples 2.2 and 2.5) of the form:

$$\begin{aligned} u_t &= ww_x + v_x, \\ v_t &= -uw_x - 3wu_x, \\ w_t &= u_x. \end{aligned} \tag{2.12}$$

Internal coordinates are x, t, u_k, v_k, and w_k, where the notation is similar to (2.11) (see Example 2.8). See Example 2.13 for the total derivatives and Example 2.16 for the linearization and its adjoint.

The CDE implementation can be found in the file bou_ell1.red. It goes as follows:

```
indep_var:={x,t};
dep_var:={u,v,w};
odd_var:={p,q,r};
total_order:=5;
f_bou:={u_t - w*w_x - v_x,v_t + u*w_x + 3*w*u_x,
  w_t - u_x};
ell_function(f_bou,lbou);
adjoint_cdiffop(lbou,lbou_star);
```

The result can be checked by

```
lbou_sf(1);

p_t - q_x - r*w_x - r_x*w;

lbou_sf(2);

p*w_x + 3*p_x*w + q_t + 3*r*u_x + r_x*u;

lbou_sf(3);

- p_x + r_t;

lbou_star_sf(1);

- p_t - 2*q*w_x - 3*q_x*w + r_x;

lbou_star_sf(2);

p_x - q_t;

lbou_star_sf(3);

p_x*w + 2*q*u_x - q_x*u - r_t;
```

2.3.3 Camassa-Holm Equation

The Camassa-Holm equation, see [20],

$$\alpha(u_t + 3uu_x) - u_{txx} = 2u_x u_{xx} + uu_{xxx}, \qquad \alpha = \text{const}, \tag{2.13}$$

lives in $J^3(2,1)$, and, in the above presentation, it is not an evolution equation. A possible choice of internal coordinates is

$$u_k^i = u_{\underbrace{x\ldots x}_{i \text{ times}}\underbrace{t\ldots t}_{k \text{ times}}}, \qquad i = 0, 1, 2, \quad k = 0, 1, \ldots$$

with the total derivatives:

$$D_x = \frac{\partial}{\partial x} + \sum_{k=0}^{\infty}\left(\sum_{i=0}^{1} u_k^{i+1}\frac{\partial}{\partial u_k^i} + D_t^k(\Phi)\frac{\partial}{\partial u_k^2}\right),$$

$$D_t = \frac{\partial}{\partial t} + \sum_{k=0}^{\infty}\sum_{i=0}^{2} u_{k+1}^i\frac{\partial}{\partial u_k^i},$$

where

$$\Phi = \frac{\alpha(u_1^0 + 3u_0^0u_0^1) - u_1^2 - 2u_0^1u_0^2}{u_0^0}.$$

The linearization operator is

$$\ell_{\mathcal{E}} = \Phi - 3\alpha u_0^1 + (2u_0^2 - 3\alpha u_0^0)D_x - \alpha D_t + 2u_0^1D_x^2 + u_0^0D_x^3 + D_x^2D_t,$$

while the adjoint to it is of the form:

$$\begin{aligned}\ell_{\mathcal{E}}^* &= \Phi - 3\alpha u_0^1 - D_x \circ (2u_0^2 - 3\alpha u_0^0) + \alpha D_t + 2D_x^2 \circ u_0^1 - D_x^3 \circ u_0^0 - D_x^2D_t \\ &= (3\alpha u_0^0 - u_0^2)D_x + \alpha D_t - u_0^1D_x^2 - u_0^0D_x^3 - D_x^2D_t.\end{aligned}$$

The CDE implementation for the linearization and its adjoint can be found in the file ch_ell1.red. It goes as follows:

```
indep_var:={t,x};
dep_var:={u};
odd_var:={p};
total_order:=5;
f_ch:={alpha*(u_t + 3*u*u_x) - u_t2x
    - 2*u_x*u_2x - u*u_3x};
ell_function(f_ch,lch);
adjoint_cdiffop(lch,lch_star);
```

The result is

```
lch_sf(1);

3*alpha*p*u_x + alpha*p_t + 3*alpha*p_x*u
    - p*u_3x - 2*p_2x*u_x - p_3x*u
    - p_t2x - 2*p_x*u_2x;

lch_star_sf(1);

- alpha*p_t - 3*alpha*p_x*u + p_2x*u_x + p_3x*u
    + p_t2x + p_x*u_2x;
```

It is then necessary to restrict the above operators on the equation manifold; to this aim we can write another CDE program (in the file ch_ell2.red) introducing the CH equation:

```
indep_var:={t,x};
dep_var:={u};
odd_var:={p};
total_order:=5;
```

```
principal_der:={u_3x};
de:={(alpha*(u_t + 3*u*u_x) - u_t2x - 2*u_x*u_2x)/u};
```

The following code will load the operators from the previous computation and restrict them on the CH equation:

```
mk_superfun(lch_sf,1,1);
mk_superfun(lch_star_sf,1,1);
in "ch_ell1_res.red";
lch_sf(1):=restrict_to_equation(lch_sf(1));

( - alpha*p*u_t + alpha*p_t*u + 3*alpha*p_x*u**2
   + 2*p*u_2x*u_x + p*u_t2x - 2*p_2x*u*u_x - p_3x*u**2
   - p_t2x*u - 2*p_x*u*u_2x)/u;

lch_star_sf(1):=restrict_to_equation(lch_star_sf(1));

 - alpha*p_t - 3*alpha*p_x*u + p_2x*u_x + p_3x*u
   + p_t2x + p_x*u_2x;
```

and if one then wishes to convert the above superfunctions into $\mathscr{C}$-differential operators:

```
conv_superfun2cdiff(lch_sf,lch);
conv_superfun2cdiff(lch_star_sf,lch_star);
```

The command `restrict_to_equation` applies two Reduce rule lists for the replacement of principal derivatives by the corresponding expression in terms of parametric derivatives. The list of even principal derivatives is `repprincparam_der`, and the list of odd principal derivatives is `repprincparam_odd`.

2.3.4 *Multi-dimensional Examples*

Let us consider some examples with more than two independent variables. Such examples are usually called *multidimensional*.

2.3.4.1 Kadomtsev-Petviashvili Equation

Consider the Kadomtsev-Petviashvili (KP) equation, see [53], in the form:

$$u_{yy} = u_{tx} - u_x^2 - uu_{xx} - \frac{1}{12}u_{xxxx}. \tag{2.14}$$

For internal coordinates on $\mathscr{E}$ one can choose the functions:

$$u^{(0)}_{i,j} = u_{\underbrace{x\dots x}_{i\text{ times}}\underbrace{t\dots t}_{j\text{ times}}}, \qquad u^{(1)}_{i,j} = u_{y\underbrace{x\dots x}_{i\text{ times}}\underbrace{t\dots t}_{j\text{ times}}},$$

where $i, j \geq 0$. The total derivatives acquire the form:

$$D_x = \frac{\partial}{\partial x} + \sum_{i,j}\left(u^{(0)}_{i+1,j}\frac{\partial}{\partial u^{(0)}_{i,j}} + u^{(1)}_{i+1,j}\frac{\partial}{\partial u^{(1)}_{i,j}}\right),$$

$$D_t = \frac{\partial}{\partial t} + \sum_{i,j}\left(u^{(0)}_{i,j+1}\frac{\partial}{\partial u^{(0)}_{i,j}} + u^{(1)}_{i,j+1}\frac{\partial}{\partial u^{(1)}_{i,j}}\right),$$

$$D_y = \frac{\partial}{\partial y} + \sum_{i,j} u^{(1)}_{i,j}\frac{\partial}{\partial u^{(0)}_{i,j}} + \sum_{i,j} D^i_x D^j_t\left(u^{(0)}_{1,1} - (u^{(0)}_{1,0})^2 - u^{(0)}_{0,0}u^{(0)}_{2,0} - \frac{1}{12}u^{(0)}_{4,0}\right)\frac{\partial}{\partial u^{(1)}_{i,j}}$$

in these coordinates. The linearization operator is

$$\ell_{\mathcal{E}} = D^2_y - D_x D_t + 2u_x D_x + u D^2_x + \frac{1}{12}D^4_x + u_{xx},$$

while its adjoint takes the form:

$$\ell^*_{\mathcal{E}} = D^2_y - D_x D_t - 2D_x \circ u_x + D^2_x \circ u + \frac{1}{12}D^4_x + u_{xx} = D^2_y - D_x D_t + u D^2_x + \frac{1}{12}D^4_x.$$

The linearization and its adjoint are computed in the example file kp_ell1.red:

```
indep_var:={t,x,y};
dep_var:={u};
odd_var:={p};
total_order:=6;
% We omit the cde call
f_kp:={u_2y - (u_tx-u_x**2-u*u_2x-(1/12)*u_4x)};

% Linearization and its adjoint
ell_function(f_kp,lkp);
adjoint_cdiffop(lkp,lkp_star);
```

with results

```
lkp_sf(1);

(12*p*u_2x + 12*p_2x*u + 12*p_2y + p_4x
 - 12*p_tx + 24*p_x*u_x)/12;

lkp_star_sf(1);
```

```
(12*p_2x*u + 12*p_2y + p_4x - 12*p_tx)/12;
```

The equation admits an evolutionary form[9] (the variable y formally playing the role of "time"):

$$u_y = v_x, \qquad v_y = u_t - uu_x - \frac{1}{12}u_{xxx}. \tag{2.15}$$

A "natural" choice of internal coordinates is

$$u_{i,j} = u_{\underbrace{x\ldots x}_{i\text{ times}}\underbrace{t\ldots t}_{j\text{ times}}}, \qquad v_{i,j} = v_{\underbrace{x\ldots x}_{i\text{ times}}\underbrace{t\ldots t}_{j\text{ times}}}$$

in this case. Then the total derivatives are

$$D_x = \frac{\partial}{\partial x} + \sum_{i,j}\left(u_{i+1,j}\frac{\partial}{\partial u_{i,j}} + v_{i+1,j}\frac{\partial}{\partial v_{i,j}}\right),$$

$$D_t = \frac{\partial}{\partial t} + \sum_{i,j}\left(u_{i,j+1}\frac{\partial}{\partial u_{i,j}} + v_{i,j+1}\frac{\partial}{\partial v_{i,j}}\right),$$

$$D_y = \frac{\partial}{\partial y} + \sum_{i,j} v_{i+1,j}\frac{\partial}{\partial u_{i,j}} + \sum_{i,j} D_x^i D_t^j\left(u_{0,1} - u_{0,0}u_{1,0} - \frac{1}{12}u_{3,0}\right)\frac{\partial}{\partial u_{i,j}}$$

The linearization and its adjoint are

$$\ell_{\mathscr{E}} = \begin{pmatrix} D_y & -D_x \\ -D_t + uD_x + u_x + \frac{1}{12}D_x^3 & D_y \end{pmatrix}$$

and

$$\ell_{\mathscr{E}}^* = \begin{pmatrix} -D_y & D_t - D_x \circ u + u_x - \frac{1}{12}D_x^3 \\ D_x & -D_y \end{pmatrix} = \begin{pmatrix} -D_y & D_t - uD_x - \frac{1}{12}D_x^3 \\ D_x & -D_y \end{pmatrix}$$

respectively, for this choice.

Then linearization and adjoint are computed in the example file kpev_ell1.red as above

```
f_kpev:={u_y - v_x,v_y - (u_t - u*u_x - (1/12)*u_3x)};
ell_function(f_kpev,lkpev);
adjoint_cdiffop(lkpev,lkpev_star);
```

[9] Actually, this transformation of the KP equation to the evolutionary form is nonlocal. See Chap. 3 for a detailed discussion.

and the results are

```
lkpev_sf(1);

p_y - q_x;

lkpev_sf(2);

(12*p*u_x + p_3x - 12*p_t + 12*p_x*u + 12*q_y)/12;

lkpev_star_sf(1);

( - 12*p_y - q_3x + 12*q_t - 12*q_x*u)/12;

lkpev_star_sf(2);

p_x - q_y;
```

2.3.4.2 Plebanski Equation

Let us consider the Plebanski (or second Heavenly) equation, see [115],

$$F = u_{tt}u_{xx} - u_{tx}^2 + u_{xz} + u_{ty} = 0. \tag{2.16}$$

Choose the functions:

$$u^{(0)}_{i,j} = u_{\underbrace{x\ldots x}_{i\text{ times}}\underbrace{z\ldots z}_{j\text{ times}}}, \quad u^{(t)}_{i,j,k} = u_{\underbrace{x\ldots x}_{i\text{ times}}\underbrace{z\ldots z}_{j\text{ times}}\underbrace{t\ldots t}_{k\text{ times}}}, \quad u^{(y)}_{i,j,k} = u_{\underbrace{x\ldots x}_{i\text{ times}}\underbrace{z\ldots z}_{j\text{ times}}\underbrace{y\ldots y}_{k\text{ times}}}$$

for internal coordinates, where $i, j \geq 0$, $k \geq 1$. The total derivatives are

$$D_x = \frac{\partial}{\partial x} + \sum_{i,j} u^{(0)}_{i+1,j} \frac{\partial}{\partial u^{(0)}_{i,j}} + \sum_{i,j,k} \left(u^{(t)}_{i+1,j,k} \frac{\partial}{\partial u^{(t)}_{i,j,k}} + u^{(y)}_{i+1,j,k} \frac{\partial}{\partial u^{(y)}_{i,j,k}} \right),$$

$$D_z = \frac{\partial}{\partial z} + \sum_{i,j} u^{(0)}_{i,j+1} \frac{\partial}{\partial u^{(0)}_{i,j}} + \sum_{i,j,k} \left(u^{(t)}_{i,j+1,k} \frac{\partial}{\partial u^{(t)}_{i,j,k}} + u^{(y)}_{i,j+1,k} \frac{\partial}{\partial u^{(y)}_{i,j,k}} \right),$$

$$D_y = \frac{\partial}{\partial y} + \sum_{i,j} u^{(y)}_{i,j,1} \frac{\partial}{\partial u^{(0)}_{i,j}} + \sum_{i,j,k} u^{(y)}_{i,j,k+1} \frac{\partial}{\partial u^{(y)}_{i,j,k}}$$
$$+ \sum_{i,j,k} D_x^i D_z^j D_t^{k-1} \left(\left(u^{(t)}_{1,0,1} \right)^2 - u^{(t)}_{0,0,2} u^{(0)}_{2,0} - u^{(0)}_{1,1} \right) \frac{\partial}{\partial u^{(t)}_{i,j,k}},$$

$$D_t = \frac{\partial}{\partial t} + \sum_{i,j} u^{(t)}_{i,j,1} \frac{\partial}{\partial u^{(0)}_{i,j}} + \sum_{i,j,k} u^{(t)}_{i,j,k+1} \frac{\partial}{\partial u^{(t)}_{i,j,k}}$$
$$+ \sum_{i,j,k} D^i_x D^j_z D^{k-1}_y \left(\left(u^{(t)}_{1,0,1}\right)^2 - u^{(t)}_{0,0,2} u^{(0)}_{2,0} - u^{(0)}_{1,1} \right) \frac{\partial}{\partial u^{(y)}_{i,j,k}}$$

in these coordinates. Linearization of Eq. (2.16) is

$$\ell_F = u_{xx} D_t^2 + u_{tt} D_x^2 - 2u_{tx} D_x D_t + D_x D_z + D_t D_y,$$

while its adjoint if of the form:

$$\ell^*_F = D_t^2 \circ u_{xx} + D_x^2 \circ u_{tt} - 2D_x D_t \circ u_{tx} + D_x D_z + D_t D_y$$
$$= u_{xx} D_t^2 + u_{tt} D_x^2 - 2u_{tx} D_x D_t + D_x D_z + D_t D_y.$$

Let us now compute linearization and its adjoint by means of CDE; the program file is ple_ell1.red. The jet space is generated as follows:

```
% Initialization of the jet environment
% of the differential equation.
indep_var:={t,x,y,z};
dep_var:={u};
odd_var:={p};
total_order:=6;
cde({indep_var,dep_var,odd_var,total_order},
   {principal_der,de,principal_odd,de_odd});
```

Let us introduce the function that defines the equation:

```
f_ple:={u_xz +u_ty - u_tx**2 + u_2t*u_2x};
```

The linearization and its adjoint are readily computed:

```
ell_function(f_ple,lple);
adjoint_cdiffop(lple,lple_star);
```

with the following results:

```
lple_sf(1);

p_2t*u_2x + p_2x*u_2t - 2*p_tx*u_tx  + p_ty + p_xz;

lple_star_sf(1);

p_2t*u_2x + p_2x*u_2t - 2*p_tx*u_tx  + p_ty + p_xz;
```

We stress that the linearization of the Plebanski equation coincides with its adjoint. Hence, due to Remark 1.3, this fact is equivalent to the equation being an Euler-Lagrange one. Indeed, the corresponding Lagrangian is

$$L = \left(\frac{1}{3}u(u_{xx}u_{tt} - u_{xt}^2) - \frac{1}{2}(u_x u_z - u_t u_y)\right) dx \wedge dz \wedge dt \wedge dy.$$

Chapter 3
Conservation Laws and Nonlocal Variables

Abstract We discuss here the notion of *conservation laws* and briefly the theory of *Abelian coverings* over infinitely prolonged equations. Computation of conservation laws is also closely related to that of *cosymmetries*, and we shall continue this discussion in Chap. 4 below. In this chapter we give the solution to Problems 1.7, 1.13, and 1.15 posed in Chap. 1.

3.1 General Theory

Let $\mathscr{E} \subset J^\infty(n, m)$ be an equation. A *horizontal differential i-form* on $\mathscr{E}$ is the expression

$$\omega = \sum_{\alpha_1 < \cdots < \alpha_i} a_{\alpha_1 \ldots \alpha_i} \, dx^{\alpha_1} \wedge \cdots \wedge dx^{\alpha_i},$$

where $a_{\alpha_1 \ldots \alpha_i} \in \mathscr{F}(\mathscr{E})$ for all indices $\alpha_1 < \cdots < \alpha_i$. The space of such forms is denoted by $\Lambda_h^i(\mathscr{E})$. In particular, $\Lambda^0(\mathscr{E}) = \mathscr{F}(\mathscr{E})$. The *horizontal de Rham differential* $d_h \colon \Lambda_h^i(\mathscr{E}) \to \Lambda_h^{i+1}(\mathscr{E})$ acts on ω by the formula

$$d_h\omega = \sum_{\alpha_1 < \cdots < \alpha_i} \sum_{l=1}^{n} D_l(a_{\alpha_1 \ldots \alpha_i}) \, dx^l \wedge dx^{\alpha_1} \wedge \cdots \wedge dx^{\alpha_i}.$$

Forms ω such that $d_h\omega = 0$ are called closed, forms $d_h\theta$ are exact. Any exact form is closed due to the identity $d_h \circ d_h = 0$.

3.1.1 Conservation Laws

A *conservation law* on $\mathscr{E}$ is a closed form $\omega \in \Lambda_h^{n-1}(\mathscr{E})$. A conservation law is said to be *trivial* if ω is exact; it is *nontrivial* otherwise. Two conservation laws are

J. Krasil'shchik et al., *The Symbolic Computation of Integrability Structures for Partial Differential Equations*, Texts and Monographs in Symbolic Computation,
https://doi.org/10.1007/978-3-319-71655-8_3

equivalent if their difference is a trivial one. Conservation laws $\omega_1, \dots, \omega_k$ are called *dependent* if there exist $\lambda_1, \dots, \lambda_k \in \mathbb{R}$ such that the linear combination

$$\lambda_1\omega_1 + \dots + \lambda_k\omega_k, \qquad \sum_i \lambda_i^2 \neq 0,$$

is a trivial conservation law. Otherwise, they are *independent*.

Remark 3.1 As it was already noted, strictly speaking, a conservation law is the equivalence class of a d_h-closed horizontal form ω modulo exact forms. But since in practice we work with the forms and not with their classes, we informally use this term for a particular representative of the class.

Thus, any conservation law

$$\omega = \sum_{i=1}^{n} a_i\, dx^1 \wedge \dots \wedge \widehat{dx^i} \wedge \dots \wedge dx^n$$

(the hat, as before, marks the omitted factors) is determined by a collection of functions $(a_1, \dots, a_n)$ satisfying

$$D_1(a_1) - D_2(a_2) + \dots + (-1)^n D_n(a_n) = 0.$$

Triviality means existence of functions a_{ij}, $1 \le i < j \le n$, such that

$$a_i = \sum_{j=1}^{i-1}(-1)^{j-1}D_j(a_{ji}) + \sum_{j=i+1}^{n}(-1)^{j-1}D_j(a_{ij}).$$

In particular, in the case $n = 2$ conservation laws are forms $\omega = a_1\, dx^1 + a_2\, dx^2$ satisfying the equation

$$D_1(a_2) = D_2(a_1), \tag{3.1}$$

while triviality is the existence of a *potential* a such that

$$a_1 = D_1(a), \qquad a_2 = D_2(a).$$

Example 3.1 Consider the Burgers equation

$$u_t = uu_x + u_{xx}$$

again. It can be readily rewritten as

$$D_t(u) = D_x\left(\frac{1}{2}u^2 + u_x\right)$$

and thus the form

$$\omega = u\,dx + \left(\frac{1}{2}u^2 + u_x\right) dt$$

is a conservation law which is nontrivial obviously. This is the sole nontrivial conservation law of the Burgers equation.[1]

Example 3.2 Consider the Gibbons-Tsarev equation, [41]

$$u_{yy} + u_x u_{xy} - u_y u_{xx} + 1 = 0.$$

It admits six conservation laws $\omega_i = P_i\,dx + Q_i\,dy$, $i = 0, \ldots, 5$, of first order,

$$P_0 = u_x^2 + u_y + y, \qquad Q_0 = u_x u_y,$$

$$P_1 = u_x^3 + 2u_x u_y - x, \qquad Q_1 = u_x^2 u_y + u_y^2 - 2u,$$

$$P_2 = u_x^4 + 3u_x^2 u_y + 3yu_x^2 + u_y^2 - xu_x + 3yu_y - 2u, \qquad Q_2 = u_x^3 u_y + 2u_x u_y^2 + 3yu_x u_y - xu_y - 3xy,$$

$$P_3 = u_x^5 + 4u_x^3 u_y + 4yu_x^3 + 3u_x u_y^2 + 2xu_x^2 + 8yu_x u_y - 2uu_x + 2xu_y - 4xy,$$

$$Q_3 = u_x^4 u_y + 3u_x^2 u_y^2 + 4yu_x^2 u_y + u_y^3 + 2xu_x u_y + 4yu_y^2 - 2uu_y - 8yu - 3x^2,$$

$$P_4 = u_x^6 + 5u_x^4 u_y + 5yu_x^4 + 6u_x^2 u_y^2 + 3xu_x^3 + 15yu_x^2 u_y + u_y^3 + \left(u + \frac{15}{2}y^2\right)(u_x^2 + u_y) + 6xu_x u_y + 5yu_y^2 - 5yu - 4x^2,$$

$$Q_4 = u_x^5 u_y + 4u_x^3 u_y^2 + 5yu_x^3 u_y + 3u_x u_y^3 + 3xu_x^2 u_y + 10yu_x u_y^2 + \left(u + \frac{15}{2}y^2\right) u_x u_y + 3xu_y^2 - \frac{3}{2}x\left(4u + 5y^2\right),$$

[1] In [118, 119], one can find dimension estimates for the spaces of conservation laws of even-order evolution equations.

$$P_5 = u_x^7 + 6u_x^5u_y + 6yu_x^5 + 10u_x^3u_y^2 + 4xu_x^4 + 24yu_x^3u_y + 4u_xu_y^3 + 4\left(\frac{1}{2}u + 3y^2\right)u_x^3 + 12xu_x^2u_y + 18yu_xu_y^2 + 12xyu_x^2 + 4\left(u + 6y^2\right)u_xu_y + 4xu_y^2 + 12xyu_y - 4xu,$$

$$Q_5 = u_x^6u_y + 5u_x^4u_y^2 + 6yu_x^4u_y + 6u_x^2u_y^3 + 4xu_x^3u_y + 18yu_x^2u_y^2 + u_y^4 + 4\left(\frac{1}{2}u + 3y^2\right)u_x^2u_y + 8xu_xu_y^2 + 6yu_y^3 + 12xyu_xu_y + 4\left(\frac{1}{2}u + 3y^2\right)u_y^2 - 2u^2 - 24y^2u - 6x^2y$$

and one, $\omega_{-5} = P_{-1}\,dx + Q_{-5}\,dy$, of third order,

$$P_{-5} = u_yu_{xxx} - \frac{1}{2}u_xu_{xx}^2, \qquad Q_{-5} = u_yu_{xxy} + \frac{1}{2}u_yu_{xx}^2 - u_xu_{xx}u_{xy} - \frac{1}{2}u_{xy}^2 - u_{xx}.$$

These conservation laws are independent.

Example 3.3 The universal hierarchy equation

$$u_{yy} = u_tu_{xy} - u_yu_{xt},$$

see Example 2.3, admits a *two-component conservation law*

$$\omega = \frac{u_t}{u_y}dx \wedge dt + \frac{1}{u_y}dy \wedge dt,$$

as well as

$$\theta = u_xu_y\,dx \wedge dt + (u_xu_t - u_y)\,dx \wedge dy,$$

and they are independent (see [14]).

Example 3.4 The Kadomtsev-Petviashvili (KP) equation (2.14)

$$u_{yy} = u_{tx} - u_x^2 - uu_{xx} - \frac{1}{12}u_{xxxx}$$

may be rewritten as

$$D_y(u_y) = D_x\left(u_t - uu_x - \frac{1}{12}u_{xxx}\right),$$

and thus the form

$$\omega = u_y\, dx \wedge dt + \left(u_t - uu_x - \frac{1}{12}u_{xxx}\right) dy \wedge dt$$

is a two-component conservation law of this equation.

Let ω be a conservation law and let us expand the form ω from $\mathcal{E}$ to $J^\infty(n, m)$ in an arbitrary way. Then due to the regularity condition, one has

$$D_1(a_1) - D_2(a_2) + \cdots + (-1)^n D_n(a_n) = \Delta(F) \tag{3.2}$$

for some $\mathcal{C}$-differential operator $\Delta\colon \mathcal{F}(n, m; r) \to \Lambda_h^n(J^k(n, m))$. The function $\psi_\omega = \Delta^*(1)|_{\mathcal{E}}$ is called the *generating function* of the conservation law ω. Generating functions provide an efficient method to compute conservation laws discussed below in Chap. 4.

3.1.2 Nonlocal Variables

Let

$$\omega = X_1\, dx^1 \wedge dx^3 \wedge \cdots \wedge dx^n + X_2\, dx^2 \wedge dx^3 \wedge \cdots \wedge dx^n,$$

be a two-component conservation law.

Remark 3.2 Of course, all conservation laws are two-component in the two-dimensional case.

Extend the ambient jet space by new unknown functions w_σ, where σ is a symmetric multi-index consisting of integers $3, \dots, n$, and consider in this extension the equation $\mathcal{E}$ together with the new relations:

$$\frac{\partial w^\sigma}{\partial x^1} = D_\sigma(X_1), \quad \frac{\partial w^\sigma}{\partial x^2} = D_\sigma(X_2), \quad \frac{\partial w^\sigma}{\partial x^i} = w^{\sigma i}, \quad i > 2, \tag{3.3}$$

where, for the convenience of notation, we relabeled $X_1 = a_2$ and $X_2 = a_1$. Equations (3.3) are compatible modulo $\mathcal{E}$. The obtained equation $\tilde{\mathcal{E}}$ in the unknowns u and w is called the Abelian covering of $\mathcal{E}$ associated to the conservation law ω. The new variables w^σ are called *nonlocal*.

Remark 3.3 In the two-dimensional case, only one nonlocal variable $w = w^{\varnothing}$ arises, while in the multi-dimensional situation, the number of new nonlocal variables is infinite.

The total derivatives on $\tilde{\mathscr{E}}$ are

$$\tilde{D}_i = D_i + \sum_{\sigma} D_\sigma(X_i)\frac{\partial}{\partial w^\sigma}, \quad i = 1, 2 \qquad \tilde{D}_i = D_i + \sum_{\sigma} w^{\sigma i}\frac{\partial}{\partial w^\sigma}, \quad i > 2,$$

where D_i are the total derivatives on $\mathscr{E}$.

Remark 3.4 Any $\mathscr{C}$-differential operator $\Delta = (\sum_\sigma a^j_{\sigma,i} D_\sigma)$ on $\mathscr{E}$ is naturally lifted to a $\mathscr{C}$-differential operator $\tilde{\Delta} = (\sum_\sigma a^j_{\sigma,i}\tilde{D}_\sigma)$ on $\tilde{\mathscr{E}}$.

Remark 3.5 All constructions accomplished on $\tilde{\mathscr{E}}$ are said to be nonlocal for the equation $\mathscr{E}$. For example, conservation laws of $\tilde{\mathscr{E}}$ are nonlocal conservation laws of $\mathscr{E}$ if they depend on the nonlocal variables w^σ. The same is valid for cosymmetries (Chap. 4), symmetries (Chap. 5), etc.

Remark 3.6 The above considered coverings constitute a particular case of a more general construction. Let $\mathscr{E} \subset J^\infty(n, m)$ be an equation in unknowns $u^1, \dots, u^m$. Introduce additional unknown functions $w^1, \dots, w^j, \dots$ and add to $\mathscr{E}$ the conditions

$$\frac{\partial w^j}{\partial x^i} = X^j_i, \tag{3.4}$$

where X^j_i are functions in x, u_σ, and w. The resulting system $\tilde{\mathscr{E}}$ is a *covering* over $\mathscr{E}$ if (3.4) is compatible modulo $\mathscr{E}$ (see also Sect. 1.1).

As it was mentioned above, coverings constructed by means of conservation laws are called *Abelian*. The following criterion can be used to check whether a covering is Abelian:

Proposition 3.1 *Let $\tau\colon \tilde{\mathscr{E}} \to \mathscr{E}$ be a finite-dimensional covering of rank r over $\mathscr{E}$. The covering is Abelian if and only if there exist r independent conservation laws $\omega_1, \dots, \omega_r$ on $\mathscr{E}$ such that the horizontal forms $\tau^*(\omega_i)$ are exact on $\tilde{\mathscr{E}}$.*

Example 3.5 Consider the covering over the Burgers equation associated with the conservation law from Example 3.1. Let w be the nonlocal variable in this covering. Then

$$w_x = u, \qquad w_t = \frac{1}{2}u^2 + u_x.$$

Hence, the total derivatives on the covering equation are

$$\tilde{D}_x = D_x + u\frac{\partial}{\partial w}, \qquad \tilde{D}_t = D_t + \left(\frac{1}{2}u^2 + u_x\right)\frac{\partial}{\partial w}.$$

This means that the covering equation is

$$w_t = \frac{1}{2}w_x^2 + w_{xx},$$

which by the change of variables $w = 2\ln|v|$ transforms to

$$v_t = v_{xx},$$

i.e., to the linear heat equation. In other words, the covering

$$v_x = \frac{1}{2}vu, \qquad v_t = \frac{1}{2}\left(\frac{1}{2}u^2 + u_x\right)$$

is a covering of the heat equation over the Burgers one. The corresponding differential substitution

$$u = \frac{2v_x}{v}$$

is called the *Cole-Hopf transformation*.

Example 3.6 Let ω be the conservation law of the KP equation, considered in Example 3.4. The covering, defined by ω is

$$v_x = u_y, \qquad v_y = \left(u_t - uu_x - \frac{1}{12}u_{xxx}\right).$$

This system coincides with the KP equation in evolutionary form (Sect. 2.3.4.1).

Example 3.7 Consider the system

$$w_y + vw_x = \frac{1}{v-w}, \qquad v_y + wv_x = \frac{1}{w-v},$$

which obviously possesses the conservation law

$$\omega = (v+w)\,dx - vw\,dy.$$

Then the defining equations for the associated Abelian covering are

$$u_x = v + w, \qquad u_y = -vw.$$

It is readily checked that the covering equation is the Gibbons-Tsarev equation (see Example 3.2).

Example 3.8 The equation of S-deformable surfaces (see [30]) is a two-component system of the form

$$u_{yy} = \frac{\frac{1}{2}\eta_1' v^2 v_x + \frac{1}{2}\eta_2' uvu_y + (\eta_1 - \eta_2)uu_y v_y + u^2 v^3}{uv(\eta_1 - \eta_2)},$$

$$v_{xx} = \frac{\frac{1}{2}\eta_1' uvv_x + \frac{1}{2}\eta_2' u^2 u_y + (\eta_2 - \eta_1)vv_x u_x + u^3 v^2}{uv(\eta_2 - \eta_1)},$$

where $\eta_1 = \eta_1(x)$, $\eta_2 = \eta_2(y)$, and "prime" denotes the corresponding derivative. The system admits, in particular, the following conservation law (see [77]):

$$\omega = \frac{u_y}{v}\,dx - \frac{v_x}{u}\,dy$$

with the corresponding covering $\tau\colon \tilde{\mathscr{E}} \to \mathscr{E}$ given by the equations

$$w_x = \frac{u_y}{v}, \qquad w_y = -\frac{v_x}{u}.$$

It can be also checked that two new conservation laws arise on $\tilde{\mathscr{E}}$:

$$\omega_1 = u \cos w\,dx + v \sin w\,dy \quad \text{and} \quad \omega_2 = -u \sin w\,dx + v \cos w\,dy.$$

A conservation law of the covering equation $\tilde{\mathscr{E}}$ that is not of the form $\tau^*(\omega)$, where ω is a conservation law of $\mathscr{E}$, is called *nonlocal*, as it was already mentioned before. Thus, ω_1 and ω_2 are nonlocal conservation laws of the equation of S-deformable surfaces.

Remark 3.7 Coming back to Example 3.5, note that the heat equation $v_t = v_{xx}$ obviously admits a conservation law $\omega = v\,dx + v_x\,dt$, which is a nonlocal conservation law of the Burgers equation.

Example 3.9 The 3D system

$$Au_t u_{xy} + Bu_x u_{ty} + Cu_y u_{tx} = 0,$$

where $A, B, C \in \mathbb{R}$, is called the *ABC*-equation, see [153]. We consider the case $A + B + C \neq 0$ here (see details in [76]).

Remark 3.8 In the particular case $A = B = C$, the equation admits the Lagrangian density $L = u_x u_y u_t\,dx\,dy\,dt$.

Without loss of generality, one can assume $A \neq 0$ and introduce the notation

$$\varkappa_1 = -\frac{B}{A}, \quad \varkappa_2 = -\frac{C}{A} \qquad \varkappa_1 + \varkappa_2 \neq 1.$$

Then

$$u_{xy} = \frac{\varkappa_1 u_x u_{ty} + \varkappa_2 u_y u_{tx}}{u_t}$$

and the functions

$$u_i^{(0)} = u_{\underbrace{t\dots t}_{i \text{ times}}}, \quad u_{i,j}^{(x)} = u_{\underbrace{t\dots t}_{i \text{ times}}\underbrace{x\dots x}_{j \text{ times}}}, \quad u_{i,j}^{(y)} = u_{\underbrace{t\dots t}_{i \text{ times}}\underbrace{y\dots y}_{j \text{ times}}},$$

$i \geq 0, j \geq 1$, can be taken for internal coordinates on $\mathscr{E}$. The total derivatives are

$$D_t = \frac{\partial}{\partial t} - \sum_{i,j}\left(u_{i+1}^{(0)}\frac{\partial}{\partial u_i^{(0)}} + u_{i+1,j}^{(x)}\frac{\partial}{\partial u_{i,j}^{(x)}} + u_{i+1,j}^{(y)}\frac{\partial}{\partial u_{i,j}^{(y)}}\right),$$

$$D_x = \frac{\partial}{\partial x} - \sum_i u_{i,1}^{(x)}\frac{\partial}{\partial u_i^{(0)}} + \sum_{i,j} u_{i,j+1}^{(x)}\frac{\partial}{\partial u_{i,j}^{(x)}} + \sum_{i,j} D_t^i D_y^{j-1}\left(\frac{\varkappa_1 u_x u_{ty} + \varkappa_2 u_y u_{tx}}{u_t}\right)\frac{\partial}{\partial u_{i,j}^{(y)}},$$

$$D_y = \frac{\partial}{\partial y} + \sum_i u_{i,1}^{(y)}\frac{\partial}{\partial u_i^{(0)}} + \sum_{i,j} u_{i,j+1}^{(y)}\frac{\partial}{\partial u_{i,j}^{(y)}} + \sum_{i,j} D_t^i D_x^{j-1}\left(\frac{\varkappa_1 u_x u_{ty} + \varkappa_2 u_y u_{tx}}{u_t}\right)\frac{\partial}{\partial u_{i,j}^{(x)}}$$

in these coordinates.

It can be checked that the equation admits the following non-Abelian covering

$$\begin{aligned} w_t &= -\frac{u_x^{(\varkappa_2-\varkappa_1-1)/\varkappa_1} w\big(\varkappa_1 u_x u_t w_x + (\varkappa_1 + \varkappa_2 - 1)w(u_t u_{xx} - \varkappa_1 u_x u_{xt})\big)}{\varkappa_1^2(\varkappa_1 + \varkappa_2 - 1)}, \\ w_y &= -\frac{\varkappa_2 u_x^{(\varkappa_2-\varkappa_1-1)/\varkappa_1} w u_y\big(\varkappa_1 u_x w_x + (\varkappa_1 + \varkappa_2 - 1)w u_{xx}\big)}{\varkappa_1^2\big(u_x^{(\varkappa_1+\varkappa_2-1)/\varkappa_1} w + \varkappa_1 + \varkappa_2 - 1\big)}, \end{aligned} \tag{3.5}$$

where w is a nonlocal variable.

Remark 3.9 Actually, we must introduce infinite number of nonlocal variables

$$w^i = w_{\underbrace{x\dots x}_{i \text{ times}}}$$

and extend (3.5) as it was indicated in Eq. (3.3).

We now apply to (3.5) the procedure of the so-called *Pavlov eversion* (see [113] and also [69] for geometric interpretation) and obtain the following Lax pair

$$\Psi_t = \frac{\lambda u_x^{(\varkappa_2-\varkappa_1-1)/\varkappa_1}(-\varkappa_1 u_x u_t \Psi_x + (\varkappa_1+\varkappa_2-1)\lambda(u_t u_{xx} - \varkappa_1 u_x u_{xt})\Psi_\lambda)}{\varkappa_1^2(\varkappa_1+\varkappa_2-1)},$$

$$\Psi_y = \frac{\varkappa_2 \lambda u_x^{(\varkappa_2-\varkappa_1-1)/\varkappa_1} u_y(-\varkappa_1 u_x \Psi_x + (\varkappa_1+\varkappa_2-1)\lambda u_{xx}\Psi_\lambda)}{\varkappa_1^2(u_x^{(\varkappa_1+\varkappa_2-1)/\varkappa_1}\lambda + \varkappa_1+\varkappa_2-1)}, \tag{3.6}$$

where Ψ is a nonlocal variable and λ is a spectral parameter now.

Let us set

$$\Psi = \lambda + \sum_{j\geq 2} \lambda^j \psi_j$$

and expand System (3.6) in formal series of λ. Then one obtains

$$\psi_{j,t} = \frac{u_x^{(\varkappa_2-\varkappa_1-1)/\varkappa_1}(-\varkappa_1 u_x u_t \psi_{j-1,x} + (\varkappa_1+\varkappa_2-1)(j-1)(u_t u_{xx} - \varkappa_1 u_x u_{xt})\psi_{j-1})}{\varkappa_1^2(\varkappa_1+\varkappa_2-1)},$$

$$\psi_{j,y} = \frac{\varkappa_2 u_x^{(\varkappa_2-\varkappa_1-1)/\varkappa_1} u_y(-\varkappa_1 u_x \psi_{j-1,x} + (\varkappa_1+\varkappa_2-1)(j-1)u_{xx}\psi_{j-1})}{\varkappa_1^2(\varkappa_1+\varkappa_2-1)}$$
$$- \frac{u_x^{(\varkappa_1+\varkappa_2-1)/\varkappa_1}\psi_{j-1,y}}{\varkappa_1+\varkappa_2-1},$$

for all $j \geq 2$. For $j = 2$, we get a two-component conservation law

$$\omega_1 = \left(\frac{\varkappa_2 u_x^{(\varkappa_2-\varkappa_1-1)/\varkappa_1} u_y u_{xx}}{\varkappa_1^2}\right) dy + \left(\frac{u_x^{(\varkappa_2-\varkappa_1-1)/\varkappa_1}(u_t u_{xx} - \varkappa_1 u_x u_{xt})}{\varkappa_1^2}\right) dt$$

of the *ABC*-equation. In general, let $\mathscr{E}_k$ denote the equation defined by the first k relations in the above system. Then we obtain an infinite tower of Abelian coverings

$$\dots \longrightarrow \mathscr{E}_k \xrightarrow{\tau_k} \mathscr{E}_{k-1} \longrightarrow \dots \longrightarrow \mathscr{E}_1 \xrightarrow{\tau_1} \mathscr{E},$$

and τ_k is associated with a two-component conservation law of the equation $\mathscr{E}_{k-1}$. Thus, we obtain an infinite series of conservation laws $\omega_2, \dots, \omega_k, \dots$ for the *ABC*-equation.

Example 3.10 The universal hierarchy equation (UHE)

$$u_{yy} - u_t u_{xy} + u_y u_{xt} = 0$$

(Example 2.3) possesses the following Lax pair, see [104],

$$w_t = \lambda^{-2}(\lambda u_t - u_y)w_x, \qquad w_y = \lambda^{-1}u_y w_x.$$

Following [14], consider the expansion

$$w = \sum_{i\in\mathbb{Z}} \lambda^i w_i.$$

Then

$$w_{i,t} = u_t w_{i+1,x} - u_y w_{i+2,x}, \qquad w_{i,y} = u_y w_{i+1,x} \tag{3.7}$$

for all integer i. The obtained infinite system of relations is useless in this form, and to improve the situation, we do the following.

Let us consider two subsystems of (3.7). The first, which is called "positive," is constructed under the assumption $w_i = 0$ for all $i \leq 0$. Then, after relabeling, we obtain the following infinite-dimensional covering

$$q_{1,y} = \frac{u_t}{u_y}, \qquad q_{1,x} = \frac{1}{u_y};$$

$$q_{i,y} = \frac{u_t}{u_y}q_{i-1,y} - q_{i-1,t}, \qquad q_{i,x} = \frac{q_{i-1,y}}{u_y},$$

where $i > 1$. The first pair of equations defines a local conservation law of the UHE, while the others give us an infinite series of nonlocal ones.

Remark 3.10 Consider the above mentioned conservation law and associated Abelian covering. Due to its definition, one has[2]

$$u_y = \frac{1}{z_x}, \qquad u_t = \frac{z_y}{z_x}.$$

Thus, we obtain the compatibility condition $(u_y)_t = (u_t)_y$, or

$$z_{xt} = z_y z_{xy} - z_x z_{yy}.$$

[2]We denote q_1 by z below.

In other words, we obtained the diagram of coverings

$$\boxed{\tilde{\mathcal{E}}\colon z_y = \frac{u_t}{u_y}, \qquad z_x = \frac{1}{u_y}}$$

$$\swarrow \qquad\qquad\qquad\qquad \searrow$$

$$\boxed{\mathcal{E}_1\colon u_{yy} = u_t u_{xy} - u_y u_{xt}} \qquad\qquad \boxed{\mathcal{E}_2\colon z_{xt} = z_y z_{xy} - z_x z_{yy}},$$

i.e., a Bäcklund transformation between UHE and the equation $z_{xt} = z_y z_{xy} - z_x z_{yy}$.

To obtain the "negative" hierarchy, we set $w_i = 0$ for $i \geq 0$ in (3.7). The system we get is equivalent to

$$\begin{aligned} r_{1,y} &= u_x u_y, & r_{1,t} &= u_x u_t - u_y; \\ r_{i,y} &= u_y r_{i-1,x}, & r_{i,t} &= u_t r_{i-1,x} - r_{i-1,y}, \end{aligned}$$

$i > 1$. The first equation corresponds to the local conservation law already described in Example 3.3; the others define an infinite series of nonlocal conservation laws.

3.2 Examples

The problem of symbolic computation of nontrivial conservation laws has been considered by many authors (see, e.g., [42, 117, 148]). The problem can be split in two steps: after solving the equation $d_h\omega$ on the differential equation, one has to filter solutions in order to discard trivial conservation laws and identify solutions that differ by a trivial conservation law.

In this section, we use a *weight* approach to finding (some) local conservation laws. This approach has its roots in the following observation: *if a partial differential equation has scaling symmetries, then in many cases its integrability structures will have the same scaling symmetry.*

Concretely, if we assign weights to independent and dependent variables (even or odd) of some partial differential equation (and, consequently, to its derivative coordinates), then we can construct spaces of monomials that are homogeneous with respect to *scale transformations*:

$$(x^1, \dots, x^n, u^1, \dots, u^m) \mapsto (\lambda^{\alpha_1} x^1, \dots, \lambda^{\alpha_n} x^n, \lambda^{\beta_1} u^1, \dots, \lambda^{\beta_m} u^m)$$

where $\alpha_1, \dots, \alpha_n, \beta_1, \dots, \beta_m$ are the weights of the variables. If a differential equation is invariant with respect to a certain scale transformation, then we can look for integrability structures in spaces of homogeneous weighted polynomials.

This approach has already been used by some of us so far [43, 54–59, 61–63, 71–74]. We will make a systematic use of the weight approach in this book. As further references see also [42] and the recent review [46]. An alternative approach appeals to the concept of generating functions and is discussed in Chap. 4.

3.2.1 Korteweg-de Vries Equation

We assign to each variable (dependent or independent) an integer, the *weight*. We assume that the weight of a monomial is the sum of the weights of its factors, and that the weight of a derivative is the difference of the weight of the function and the weight of the variable:

$$[u_z] = [u] - [z], \qquad [uv] = [u] + [v]. \tag{3.8}$$

Then, we assume that the Korteweg-de Vries (KdV) equation is a sum of monomials which are homogeneous with respect to the given weights. This means that

$$[u_t] = [u_{3x}], \quad [u_{3x}] = [uu_x]. \tag{3.9}$$

In this case, as the KdV equation is a mathematical model of waves in shallow water, the variables can be given physical dimensions, and the scale symmetries acquire a physical meaning. A solution of the equation (3.9) is

$$[x] = -1, \quad [t] = -3, \quad [u] = 2$$

and set $[u_k] = 2 + k$.

Remark 3.11 In all our examples, weights of dependent variables are positive, and weights of independent variables are negative. This implies that derivative coordinates will have positive weight, thus avoiding that variables in the system have negative weights. Negative weights can yield infinite-dimensional spaces of weighted monomials of a given positive weight. This would make the search for solutions much more complicated and less efficient.

In order to solve the problem with CDE, we run the program kdv_lcl1.red with the following initial data:

```
indep_var:={x,t};
dep_var:={u};
deg_indep_var:={-1,-3};
deg_dep_var:={2};
deg_odd_var:={0};
total_order:=10;
```

We use the new global variables

```
deg_indep_var, deg_dep_var, deg_odd_var
```

to assign weights (also called *scale degrees*) to all variables in the system. The list `deg_indep_var` should have the same number of elements of the list `indep_var`, and so on.

After entering the equation

```
principal_der:={u_t};
de:={u*u_x+u_3x};
```

and calling CDE

```
cde({indep_var,dep_var,{},total_order},
{principal_der,de,{},{}});
```

the weights are declared in the system by

```
cde_grading(deg_indep_var,deg_dep_var,deg_odd_var).
```

Then we initialize two operators, one for the equations that will determine the conservation laws and another that will indicate all unknown constants:

```
operator equ,c;
```

We also initialize a counter for the unknown constants:

```
ctel:=0;
```

After this, all variables which are generated by CDE must be assigned a weight according to the above rules (3.8). This is accomplished by

```
l_grad_mon:=der_deg_ordering(0,all_parametric_der);
```

`l_grad_mon` is a list of lists of variables; the list is ordered by weight, starting from weight 1. In our case we have

```
l_grad_mon;

{{},{u},{u_x},{u_2x},{u_3x},{u_4x},{u_5x},
{u_6x},{u_7x},{u_8x},{u_9x},{u_10x}};
```

This will be used to generate the list of homogeneous monomials of given weight to be used in the ansatz for the generating function of the conservation laws:

```
gradmon:=graded_mon(1,5,l_grad_mon);
gradmon:={1} . gradmon;
gradmon;

{{1},{},{u},{u_x},{u**2,u_2x},{u*u_x,u_3x}};
```

The variable `ansatz` contains all the above monomials:

```
ansatz:=for each el in gradmon join el;
```

We suppose that the conservation law has the form $\omega = a_x dx + a_t dt$. So, we generate two assumptions on the coefficients of the conservation law:

```
ax:=(for each el in ansatz sum (c(ctel:=ctel+1)*el));
at:=(for each el in ansatz sum (c(ctel:=ctel+1)*el));
```

Next we define the equation $d_h(\omega) = 0$, where d_h, as before, is the horizontal de Rham differential (1.3) restricted to the equation:

```
operator equ;
equ 1:=td(ax,t) - td(at,x);
```

The equation can be solved with the solver in CDIFF as follows. First of all, we collect some information for the solver, namely, the list of variables with respect to which the equation is polynomial:

```
vars:=append(indep_var,all_parametric_der);
```

and the number of initial equations `tel:=1`.

Next command comes from the package CDIFF, and it is a solver of linear algebraic sparse systems of equations. It passes the equation(s) together with their number `tel`, the vector of constants `c`, its length `ctel`, an arbitrary constant `f` that may appear in computations:

```
initialize_equations(equ,tel,{},{c,ctel,0},{f,0,0});
```

As the equation(s) is polynomial, we must collect its coefficients and equate them to zero.[3] This is achieved by the following command:

```
tel:=splitvars_opequ(equ,1,1,vars);
```

The first "1" stands for the first equation and the second "1" for the last equation of the system, which is arranged as `equ(1):=...,equ(2):=...,...`. The equations which result from the above procedure are appended to the operator `equ` after the last equation of the system. The total number of entries in `equ` is `tel`, which is passed to the equation solver:

```
put_equations_used tel;
```

Finally, we solve the system obtained after the splitting procedure:

```
for i:=2:tel do integrate_equation i;
```

We get

```
ax;

c(3)*u_x + c(2)*u + c(1);

at;
```

[3]This procedure is usually called "*splitting*."

```
(2*c(8) + 2*c(3)*u*u_x + 2*c(3)*u_3x +
c(2)*u**2 + 2*c(2)*u_2x)/2;
```

It is clear that the conservation law corresponding to `c(3)` is trivial, because it has a potential that is easy to guess; even more trivial conservation laws are those who correspond to `c(1)` and `c(8)`. Here this fact is evident; how to get rid of less evident trivialities by an "automatic" mechanism?

The above problem is solved by the program kdv_lcl2.red. The basic assumption is that if a conservation law is trivial and is weighted, its potential will be weighted too.

The input data is the same jet space and the same equation as in kdv_lcl1.red, plus the above result. An operator and a counter for the unknown constants in the potential of the trivial conserved quantity must be added:

```
operator c,cc,equ;
cctel:=0;
```

The list of weighted monomials shall be modified by adding the independent variables:

```
gradmon:=indep_var . ({1} . gradmon);
```

The ansatz for the potential of the unknown trivial conservation law is generated as follows:

```
a0:=for each el in ansatz sum
  (cc(cctel:=cctel+1)*el);
```

The equations are

```
equ 1:=ax-td(a0,x);
equ 2:=at-td(a0,t);
```

having loaded the values `ax` and `at` found by the previous program. We initialize the solver by

```
vars:=append(indep_var,all_parametric_der);
tel:=2;
initialize_equations(equ,tel,{},{cc,cctel,0},{f,0,0});
```

and solve it through splitting of the first two equations:

```
tel:=splitvars_opequ(equ,1,2,vars);
put_equations_used tel;
for i:=3:tel do integrate_equation i;
```

Note that we do not solve the first two equations directly; we only try to solve the splitted equations, i.e., the coefficients of the polynomial equations `equ(1)` and `equ(2)`

We observe that *not all* equations in the system can be solved; in the opposite case, this would mean that all the previously found conservation laws are trivial. After solving all the equations of the system that we are able to solve, we obtain

```
axnontriv;

c(2)*u;

atnontriv;

(c(2)*(u**2 + 2*u_2x))/2;
```

Note that the components of the conservation law have different weights, being multiplied by dx and dt, respectively. A more refined ansatz should take into account this property, as we will see.

3.2.2 Dispersionless Boussinesq System

Simple examples of conservation laws for Eq. (2.12) are

$$\omega_1 = w\,dx + u\,dt, \quad \omega_2 = u\,dx + \left(v + \frac{w^2}{2}\right) dt, \quad \omega_3 = (v + w^2)\,dx - uw\,dt.$$

We load the jet space as before (program file bou_lcl1.red), with the equation

```
principal_der:={u_t,v_t,w_t};
de:={w*w_x + v_x, - u*w_x - 3*w*u_x,u_x};
cde({indep_var,dep_var,odd_var,total_order},
{principal_der,de,{},{}});
```

As an ansatz for the above conservation laws, we can require that the components be quadratic in the dependent variables. The CDE code for the ansatz is

```
ansatz:=1 . mkallmon(2,dep_var);
ctel:=0;
operator c,equ;
fx:=for each el in ansatz sum c(ctel:=ctel+1)*el;
ft:=for each el in ansatz sum c(ctel:=ctel+1)*el;
```

In particular, the instruction that generates all monomials of degrees from 1 to the specified number (2 in our case) is mkallmon. Solving the equations as in the previous example yields the required conserved quantities plus two additional trivial conservation laws:

```
fx;

c(4)*u + c(3)*v + c(3)*w**2 + c(2)*w + c(1);

ft;
```

```
(2*c(11) + 2*c(4)*v + c(4)*w**2 - 2*c(3)*u*w
+ 2*c(2)*u)/2;
```

3.2.3 Camassa-Holm Equation

For Eq. (2.13), examples of conservation laws are

$$\omega_1 = \left(u_0^0 u_0^2 + \frac{u_1^2 - 3\alpha (u_0^0)^2}{2} \right) dx + (\alpha u_0^0 - u_0^2)\, dt,$$

$$\omega_2 = (-2\alpha (u_0^0)^3 + 2(u_0^0)^2 u_0^2 + u_0^0 u_1^1 - u_1^0 u_0^1)\, dx + (\alpha (u_0^0)^2 - u_0^0 u_0^2)\, dt.$$

The code is similar to what was done in Sect. 3.2.2 and is left to the reader.

3.2.4 Gibbons-Tsarev Equation

Here we describe the program gt_lcl1.red for computing the conservation laws of the Gibbons-Tsarev equation (Example 3.2).

The differential equation can be entered as

```
principal_der:={u_2y};
de:={ - u_x*u_xy + u_y*u_2x - 1};
```

After calling `cde`, let us initialize constants and equations

```
ctel:=0;
operator c,equ;
```

We look for polynomial conservation laws of degree at most 3 with respect to independent variables, dependent variables, and their first-order derivatives. The form is `cy dx + cx dy`. First of all we collect the variables:

```
top_degree:=3;
even_vars:=for i:=0:1 join
  selectvars(0,i,dep_var,all_parametric_der);
all_vars:=append(indep_var,even_vars);
```

Then we generate all monomials:

```
list_mon:=1 . mkallmon(top_degree,all_vars);
```

Let us generate linear combinations of the monomials in order to build the ansatz and the equation.

```
cx_list:=for each el in list_mon collect
  (c(ctel:=ctel+1)*el);
```

```
cy_list:=for each el in list_mon collect
  (c(ctel:=ctel+1)*el);
cx_x_list:=for each el in cx_list collect td(el,x);
cy_y_list:=for each el in cy_list collect td(el,y);
cx:=(part(cx_list,0):=plus);
cy:=(part(cy_list,0):=plus);
cx_x:=(part(cx_x_list,0):=plus);
cy_y:=(part(cy_y_list,0):=plus);
```

Remark 3.12 In the above code we generated a linear combination of many monomials in two steps: we first made a list of monomials with instructions like

```
cx_list:=for each el in list_mon collect
  (c(ctel:=ctel+1)*el);
```

then we turned it into a linear combination of its element by transforming the list into a sum:

```
cx:=(part(cx_list,0):=plus);
```

This trick has the result of invoking Reduce simplifier only once, after the second command. If one would use the standard, more direct, command

```
cx:=for each el in list_mon sum (c(ctel:=ctel+1)*el);
```

then the simplifier would be invoked at each step of the sum, and the execution speed would be much slower.

Note that by the same principle, we first compute the total derivative of each element of the list and then transform the list into a sum.

The equation

```
  equ 1:=cy_y - cx_x;
```

The solutions of the above equation must be filtered by polynomial splitting, and the resulting algebraic system on the constants `c(1)`, `c(2)`,... will be solved. The many solutions that we obtain must be filtered in order to obtain nontrivial and independent conservation laws.

As an ansatz for the potential of trivial conservation laws we can consider a polynomial function of one (algebraic) degree higher than the conservation laws. Let us initialize a counter:

```
operator cc;
cctel:=0;
```

We suppose that the trivial conservation law has the form `tcl` and its differential is `tcl_x dx + tcl_y dy` and is equal to the conservation law `cy dx + cx dy` that we found so far. Let us introduce the variables for the ansatz:

```
top_degree_pot:=top_degree+1;
even_vars_pot:=for i:=0:1 join
  selectvars(0,i,dep_var,all_parametric_der);
all_vars_pot:=append(indep_var,even_vars_pot);
```

and the monomials for the ansatz:

```
list_mon_pot:=
  1 . mkallmon(top_degree_pot,all_vars_pot);
```

Then, we generate the equations as follows:

```
tcl_list:=for each el in list_mon_pot collect
  (cc(cctel:=cctel+1)*el);
tcl_x_list:=for each el in tcl_list collect
  td(el,x);
tcl_y_list:=for each el in tcl_list collect
  td(el,y);
tcl:=(part(tcl_list,0):=plus);
tcl_x:=(part(tcl_x_list,0):=plus);
tcl_y:=(part(tcl_y_list,0):=plus);

clear equ;
operator equ;

equ 1:=tcl_x - cy;
equ 2:=tcl_y - cx;
```

Of course, if there is at least one nontrivial conservation law in the set of solutions that we found before, we will fail to solve the above system, but the trivial conservation laws will be canceled in the process. After solving the linear algebraic system in the unknowns `cc(1)`, `cc(2)`, ..., we remain with the nontrivial conservation laws:

```
cynontriv:=equ 1;
cxnontriv:=equ 2;
```

The computation does not yield a mathematical proof that the conservation laws that we obtained are nontrivial. Indeed, in principle, the conservation laws could be the differential of a nonhomogeneous one-form. However, as we will see in Chap. 4, for evolution equations it is always possible to decide (by a computation) if a conservation law is trivial or not. In particular, in our case, a computation confirms the nontriviality of the solutions and shows that all solutions lie in two different equivalence classes which are generated by the conservation laws that correspond to the two constants `c(18)` and `c(17)`:

```
df(cxnontriv,c(18));

 - u_x*u_y + x;

df(cynontriv,c(18));

- (u_x**2 + u_y);
```

```
df(cxnontriv,c(17));

2*u - u_x**2*u_y - u_y**2;

df(cynontriv,c(17));

u_x*( - u_x**2 - 2*u_y);
```

They correspond to the conservation laws ω_0 and ω_1 of Example 3.2 (up to trivial conservation laws!). It is easy to obtain higher algebraic degree conservation laws: it is sufficient to increase the value of the variable `top_degree`.

3.2.5 Multi-dimensional Examples

Some examples in dimensions >2 are considered below.

3.2.5.1 Universal Hierarchy Equation

We describe the program uh_lcl1.red for computing two-component conservation laws of the universal hierarchy equation (Example 3.10). Two-component conservation laws of multi-dimensional equations are particularly important because they allow us to introduce nonlocal variables.

After the initialization

```
indep_var:={t,x,y};
dep_var:={u};
total_order:=10;
resname:="uh_lcl1_res.red";
principal_der:={u_2y};
de:={u_t*u_xy - u_y*u_tx};
```

we look for conservation laws of the type

$$\omega = Xdt \wedge dy + Ydt \wedge dx.$$

We use the identifier `cx` for X and `cy` for Y. We would like to search for conserved quantities whose coefficients are functions of the dependent variables and their first-order derivatives. Let us make a list with such symbols:

```
even_vars:=for i:=0:1 join
  selectvars(0,i,dep_var,all_parametric_der);
```

and declare the dependency of the coefficients

```
unk:={cx,cy};
for each el in even_vars do
```

```
<<
  depend(cx,el);
  depend(cy,el)
>>;
```

The equation is

```
  total_eq:={td(cx,y) - td(cy,x)};
```

The variables with respect to which the equation is polynomial:

```
split_vars:= for i:=2:total_order join
  selectvars(0,i,dep_var,all_parametric_der);
```

We solve the problem with the Reduce package CRACK. The CRACK software is a powerful program for solving in an interactive way or in batch mode overdetermined systems of PDEs [149, 150]. This program has been used since long time for the computation of symmetries and/or conservation laws of PDEs [147, 148]. CDE aims go beyond symmetries and conservation laws; however the determining systems for IS turn out to be of the same nature of those for symmetries and conservation laws, as we will see.

```
load_package crack;
crack_results:=crack(total_eq,{},unk,split_vars);
```

The result is

```
crack_results:={{{},
{cy=( - c_11*u_y**2 - c_14*u_t + c_16*u_y**2
      + c_18*u_y)/u_y,
cx=( - c_11*u_x*u_y - c_14 - c_15*u_y
      + c_16*u_x*u_y - c_17*u_y)/u_y},
{c_11,c_15,c_14,c_16,c_17,c_18},
{}}};
```

This means that there are six conservation laws which depend on dependent variables and first-order derivatives; they are the coefficients of the constants c_11, etc.

Now, there can be trivial conservation laws in the above vector space. We can assume that the potential is of the form $\alpha = Tdt$, where T is a function in independent variables, dependent variables, and first-order derivatives. The equation can be set up as

```
total_eq:={ - td(tct,y) - df(cx,c_11),
  - td(tct,x) - df(cy,c_11)};
```

where T is represented by the identifier tct. Failure in finding solutions by the following command:

```
crack_results_triv:=crack(total_eq,{},unk,split_vars);
```

means that the conservation law is nontrivial. It turns out that there are three trivial conservation laws corresponding to the constants `c_15`, `c_17`, `c_18`. Note that it would not be possible to prove that these are trivial conservation laws without allowing T to be dependent on t and x. The reader will easily recognize one of the conservation laws of Example 3.10 among the others.

3.2.5.2 Khokhlov-Zabolotskaya Equation

Let us consider the following potential form of the dispersionless limit of the Kadomtsev-Petviashvili equation (Example 3.4):

$$v_{yy} = -v_{xx}v_x + v_{tx}. \tag{3.10}$$

The above equation is known as the (potential form of) Khokhlov-Zabolotskaya equation [152]; see also Example 2 in [131] for a (3 + 1)-dimensional integrable generalization of this equation.

It can be rewritten in evolutionary form as the following system of PDEs:

$$\begin{aligned} v_y &= w \\ w_y &= -v_{xx}v_x + v_{tx}. \end{aligned} \tag{3.11}$$

We choose internal coordinates on $\mathscr{E}$ in the standard way:

$$v_{i,j} = v_{\underbrace{x\ldots x}_{i\text{ times}}\underbrace{t\ldots t}_{j\text{ times}}}, \qquad w_{i,j} = w_{\underbrace{x\ldots x}_{i\text{ times}}\underbrace{t\ldots t}_{j\text{ times}}}$$

Then the total derivatives are

$$D_x = \frac{\partial}{\partial x} + \sum_{i,j}\left(v_{i+1,j}\frac{\partial}{\partial v_{i,j}} + w_{i+1,j}\frac{\partial}{\partial w_{i,j}}\right),$$

$$D_t = \frac{\partial}{\partial t} + \sum_{i,j}\left(v_{i,j+1}\frac{\partial}{\partial v_{i,j}} + w_{i,j+1}\frac{\partial}{\partial w_{i,j}}\right),$$

$$D_y = \frac{\partial}{\partial y} + \sum_{i,j} w_{i,j}\frac{\partial}{\partial v_{i,j}} + \sum_{i,j} D_x^i D_t^j\left(-v_{2,0}v_{1,0} + v_{1,1}\right)\frac{\partial}{\partial w_{i,j}}$$

in these coordinates.

Two conservation laws can be found using the weight approach for constructing an ansatz. More precisely, we assume that

$$[v] = 1, \quad [w] = 3, \quad [t] = -3, \quad [x] = -1, \quad [y] = -2 \tag{3.12}$$

and $\omega = c_1 dx \wedge dy + c_2 dt \wedge dy + c_3 dt \wedge dx$. This means that if we assume that the conservation law is a homogeneous polynomial with respect to the weight (including the weight of independent variables in $dx \wedge dy$, etc.), it shall be $[c_1] = s$, $[c_2] = s+2$, $[c_3] = s+1$.

We now describe the program pkz_lcl1.red. The initialization looks like

```
indep_var:={t,x,y};
dep_var:={v,w};
deg_indep_var:={-3,-1,-2};
deg_dep_var:={1,3};
total_order:=10;
principal_der:={v_y,w_y};
de:={w,v_tx - v_x*v_2x};
cde({indep_var,dep_var,{},total_order},
   {principal_der,de,{},{}});
cde_grading(deg_indep_var,deg_dep_var,{});
```

We start the computation by generating the polynomial ansatz for the conservation laws. In order to narrow our search, we assume that $d = 2$ and that the components of the conservation law depend

- on derivative coordinates of order at most equal to 3;
- on independent variables t, x, y in a polynomial way up to (algebraic) degree 3.

```
ctel:=0;
operator c,equ;
% List of variables ordered by gradings
top_degree:=10;
even_vars:=for i:=0:3 join
  selectvars(0,i,dep_var,all_parametric_der);
l_grad_der:=der_deg_ordering(0,even_vars);
% List of graded monomials of
%   weight <= top_degree
gradmon:=graded_mon(1,top_degree,l_grad_var);
% Adding zero-degree terms to variables
gradmon:={1} . gradmon;

% List of weights of the components of the
% conservation laws
deg_ct:=2;
deg_cx:=deg_ct+2;
deg_cy:=deg_ct+1;
% algebraic degree of the independent variables
deg_ind:=3;
```

The function mkallgradmon_evenind creates monomials of weights equal to the third argument with monomials of independent variables of algebraic degree less than or equal to the first argument and graded monomials in the list gradmon, which is ordered by weight.

```
grmont:=
  mkallgradmon_evenind(deg_ind,gradmon,deg_ct);
grmonx:=
  mkallgradmon_evenind(deg_ind,gradmon,deg_cx);
grmony:=
  mkallgradmon_evenind(deg_ind,gradmon,deg_cy);
```

The above commands will generate a warning: indeed, not all monomials of independent variables of degree less than or equal to 3 could be used, as their multiplication with graded monomials of dependent variables would not produce any suitable monomial to our purposes.

Then we generate an ansatz, basing on the above monomials:

```
ct_list:=for each el in grmont collect
   (c(ctel:=ctel+1)*el);
cx_list:=for each el in grmonx collect
   (c(ctel:=ctel+1)*el);
cy_list:=for each el in grmony collect
   (c(ctel:=ctel+1)*el);

ct_t_list:=for each el in ct_list collect td(el,t);
cx_x_list:=for each el in cx_list collect td(el,x);
cy_y_list:=for each el in cy_list collect td(el,y);

ct:=(part(ct_list,0):=plus);
cx:=(part(cx_list,0):=plus);
cy:=(part(cy_list,0):=plus);

ct_t:=(part(ct_t_list,0):=plus);
cx_x:=(part(cx_x_list,0):=plus);
cy_y:=(part(cy_y_list,0):=plus);
```

Here, ct is the component c_1, cx is the component c_2, and cy is the component c_3 in the conservation law ω (3.12).

Finally, the equation

```
equ 1:=ct_t-cx_x+cy_y;
```

The solution contains trivial conservation laws that must be removed.

Trivial conservation laws are of the form $d_h\alpha$, where $\alpha = Tdt + Xdx + Ydy$. In the following code, T is represented by tct, X by tcx, and Y by tcy. We generate an ansatz based on the weight approach, and we solve the resulting equations. Of course, if there is at least one nontrivial conservation law in the set of solutions that we found before, we will fail, but the trivial conservation laws will be canceled in the process.

```
operator cc;
cctel:=0;

deg_tcx:=deg_ct-2;
deg_tcy:=deg_tcx+1;
deg_tct:=deg_ct;
deg_tind:=deg_ind+1;

grmon_tcx:=
  mkallgradmon_evenind(deg_tind,gradmon,deg_tcx);
grmon_tcy:=
  mkallgradmon_evenind(deg_tind,gradmon,deg_tcy);
grmon_tct:=
  mkallgradmon_evenind(deg_tind,gradmon,deg_tct);

tcx_list:=for each el in grmon_tcx collect
  (cc(cctel:=cctel+1)*el);
tcy_list:=for each el in grmon_tcy collect
  (cc(cctel:=cctel+1)*el);
tct_list:=for each el in grmon_tct collect
  (cc(cctel:=cctel+1)*el);

tcx:=(part(tcx_list,0):=plus);
tcy:=(part(tcy_list,0):=plus);
tct:=(part(tct_list,0):=plus);

clear equ;
operator equ;

equ 1:=(td(tcy,x)-td(tcx,y)) - ct;
equ 2:=(td(tcy,t)-td(tct,y)) - cx;
equ 3:=(td(tcx,t)-td(tct,x)) - cy;
```

After removing the trivial conservation laws (this implies solving a linear sparse system of more than 11,000 algebraic equations), we obtain several solutions; they are all contained in the above equations that we rename as follows:

```
ctnontriv:=equ 1;
cxnontriv:=equ 2;
cynontriv:=equ 3;
```

The above variables are the components c_1, c_2, c_3 of the nontrivial conservation laws ω that we obtained using the ansatz that was defined so far.

In our case, a computation (based on the theory of cosymmetries, see Chap. 4) confirms the nontriviality of the solutions and shows that all solutions lie in two equivalence classes (up to trivial conservation laws) which are generated by the conservation laws that correspond to the two constants `c(1488)` and `c(681)`. It is interesting to observe that they are two-component conservation laws: ctnontriv is 0 (up to trivial conservation laws). In order to select the two other components, we can issue the following commands:

```
df(cxnontriv,c(1488));

(2*v_t - v_x**2)/2;

df(cynontriv,c(1488));

w;

df(cxnontriv,c(681));

(t*( - 6*v**2*w_2x - 6*v*v_2tx*y + 18*v*v_2x*v_tx*y
+ 12*v*v_3x*v_x**2*y - 6*v*v_t2x*v_x*y - 36*v*v_x*w_x
+ 12*v*w*w_2x*y - 12*v*w_x**2*y - 3*v_2tx*v_tx*y**3
+ 6*v_2tx*v_x*x*y - 6*v_2x*v_x**3*y + 12*v_2x*w**2*y
- 6*v_3t*x*y - v_3tx*v_x*y**3 + v_4t*y**3
+ 6*v_tx**2*x*y))/6;

df(cynontriv,c(681));

(t*( - 6*v**2*v_3x - 36*v*v_2x*v_x + 12*v*v_3x*w*y
+ 6*v*v_tx - 12*v*v_x*w_2x*y - 6*v*w_tx*y
+ 6*v_2t*x + 24*v_2x*v_x*w*y - 3*v_3t*y**2
+ 6*v_tx*w*y - 10*v_x**3 - 6*v_x**2*w_x*y - 6*w**2
- 6*w_2t*x*y + w_3t*y**3))/6;
```

For the sake of convenience, we rewrite the two results in a more readable form:

$$
\begin{aligned}
\omega_1 =& \frac{2v_t - v_x^2}{2} dt \wedge dy + w dt \wedge dx \\
\omega_2 =& \frac{1}{6} t\Big(- 6v^2 w_{2x} - 6v v_{2tx} y + 18 v v_{2x} v\_txy + 12 v v_{3x} v_x^2 y - 6 v v_{t2x} v_x y - 36 v v_x w_x \\
& + 12 v w w_{2x} y - 12 v w_x^2 y - 3 v_{2tx} v_{tx} y^3 + 6 v_{2tx} v_x x y - 6 v_{2x} v_x^3 y + 12 v_{2x} w^2 y \\
& - 6 v_{3t} x y - v_{3tx} v_x y^3 + v_{4t} y^3 + 6 v_{tx}^2 x y\Big) dt \wedge dy \\
& + \frac{1}{6} t\Big(- 6 v^2 v_{3x} - 36 v v_{2x} v_x + 12 v v_{3x} w y + 6 v v_{tx} - 12 v v_x w_{2x} y - 6 v w_{tx} y \\
& + 6 v_{2t} x + 24 v_{2x} v_x w y - 3 v_{3t} y^2 + 6 v_{tx} w y - 10 v_x^3 - 6 v_x^2 w_x y - 6 w^2 \\
& - 6 w_{2t} x y + w_{3t} y^3\Big) dt \wedge dx
\end{aligned}
$$

While the closure of ω_1 is evidently the second equation of the system, to our knowledge ω_2 is a new result.

Chapter 4
Cosymmetries

Abstract The most common use of cosymmetries is related to construction of conservation laws, because the generating functions of conservation laws are cosymmetries, but they also play an important role in the theory of the tangent (Chap. 6) and the cotangent (Chap. 9) coverings. We give the solution to Problems 1.8, 1.9, 1.10 and 1.13 in this chapter.

4.1 General Theory

A cosymmetry of an equation $\mathscr{E} = \{F^1 = \cdots = F^r = 0\} \subset J^\infty(n, m)$ is a solution $\psi \in \mathscr{F}(\mathscr{E}; r)$ of the equation:

$$\ell^*_{\mathscr{E}}(\psi) = 0. \tag{4.1}$$

The space of cosymmetries is denoted by $\operatorname{cosym}(\mathscr{E})$.

Example 4.1 Let us begin with a simple example of the Burgers equation:

$$u_t = uu_x + u_{xx}.$$

Equation (4.1) is

$$D_t(\psi) = uD_x(\psi) - D_x^2(\psi)$$

in this case, where the total derivatives are as described in Remark 2.3. It is not difficult to show that the only solution of this equation is $\psi = 1$.

Example 4.2 Cosymmetries of the Gibbons-Tsarev equation

$$u_{yy} + u_x u_{xy} - u_y u_{xx} + 1 = 0$$

J. Krasil'shchik et al., *The Symbolic Computation of Integrability Structures for Partial Differential Equations*, Texts and Monographs in Symbolic Computation,
https://doi.org/10.1007/978-3-319-71655-8_4

are solutions of the equation:

$$\ell^*_{\mathscr{E}}(\psi) = 0, \tag{4.2}$$

where the linearization operator is

$$\ell_{\mathscr{E}} = D_y^2 + u_x D_x D_y - u_y D_x^2 + u_{xy} D_x - u_{xx} D_y,$$

while its adjoint is computed by

$$\begin{aligned}\ell^*_{\mathscr{E}} &= \left(D_y^2 + u_x D_x D_y - u_y D_x^2 + u_{xy} D_x - u_{xx} D_y\right)^* \\ &= D_y^2 + D_x D_y \circ u_x - D_x^2 \circ u_y + D_x \circ u_{xy} + D_y \circ u_{xx} \\ &= D_y^2 + u_x D_x D_y - u_y D_x^2 - 2u_{xy} D_x + 2u_{xx} D_y.\end{aligned}$$

The total derivatives are presented as follows. Choose the functions

$$u_k^{(0)} = u_{\underbrace{x\dots x}_{k\ \text{times}}}, \qquad u_k^{(1)} = u_{\underbrace{x\dots x}_{k\ \text{times}}y}$$

for internal coordinates. Then

$$\begin{aligned}D_x &= \frac{\partial}{\partial x} + \sum_{k\geq 0}\left(u_{k+1}^{(0)}\frac{\partial}{\partial u_k^{(0)}} + u_{k+1}^{(1)}\frac{\partial}{\partial u_k^{(1)}}\right),\\ D_y &= \frac{\partial}{\partial y} + \sum_{k\geq 0} u_k^{(1)}\frac{\partial}{\partial u_k^{(0)}} + \sum_{k\geq 0} D_x^k\left(u_0^{(1)}u_2^{(0)} - u_1^{(0)}u_1^{(1)}\right)\frac{\partial}{\partial u_k^{(1)}}.\end{aligned}$$

Then the space of solutions for Eq. (4.2) is generated by six first-order functions:

$$\begin{aligned}\psi_0 &= 1,\\ \psi_1 &= 2u_x,\\ \psi_2 &= 3u_x^2 + 2u_y + 3y,\\ \psi_3 &= 4u_x^3 + 6u_x u_y + 8yu_x + 2x,\\ \psi_4 &= 5u_x^4 + 12u_x^2 u_y + 15yu_x^2 + 3u_y^2 + 6xu_x + 10yu_y + u + \frac{15}{2}y^2,\\ \psi_5 &= 6u_x^5 + 20u_x^3 u_y + 24yu_x^3 + 12u_x u_y^2 + 12xu_x^2 + 36yu_x u_y + (4u + 24y^2)u_x\\ &\quad + 8xu_y + 12xy\end{aligned}$$

and one function

$$\psi_{-5} = u_{xxx}.$$

of third order.

Example 4.3 As it follows from Example 2.18, the adjoint linearization operator of the universal hierarchy equation is

$$\ell_{\mathcal{E}}^{*} = D_y^2 - u_t D_x D_y + u_y D_x D_t - 2u_{xt} D_y + 2u_{xy} D_t.$$

Solving the equation $\ell_{\mathcal{E}}^{*}(\psi) = 0$, we obtain the following cosymmetries:

$$\psi_1(T) = \frac{T}{u_y^2}, \qquad \psi_2(T) = \frac{u_t T}{u_y^3} - \frac{1}{2}\frac{yT'}{u_y^2},$$

$$\psi_1(X) = X, \qquad \psi_2(X) = u_x X + \frac{1}{2} u X'$$

where $T = T(t)$, $X = X(x)$ are arbitrary smooth functions and "prime" denotes the corresponding derivative.

Remark 4.1 Let $\mathcal{E}$ be an equation in unknown $u = (u^1, \dots, u^m)$ and $\tau\colon \tilde{\mathcal{E}} \to \mathcal{E}$ be a covering. Assume for simplicity that $u = u(x, y)$ and τ is one-dimensional, w being a nonlocal variable. Let the entire construction be given by the system:

$$F = 0, \qquad w_x = X, \qquad w_y = Y.$$

Then the linearization of the covering equation is

$$\ell_{\tilde{\mathcal{E}}} = \begin{pmatrix} \tilde{\ell}_{\mathcal{E}} & 0 \\ -\tilde{\ell}_X & \tilde{D}_x + \partial X/\partial w \\ -\tilde{\ell}_Y & \tilde{D}_y + \partial Y/\partial w \end{pmatrix},$$

where "tilde" denotes the lift of a $\mathcal{C}$-differential operator to the covering space. Consequently,

$$\ell_{\tilde{\mathcal{E}}}^{*} = \begin{pmatrix} \tilde{\ell}_{\mathcal{E}}^{*} & -\tilde{\ell}_X^{*} & -\tilde{\ell}_Y^{*} \\ 0 & -\tilde{D}_x + \partial X/\partial w & -\tilde{D}_y + \partial Y/\partial w \end{pmatrix}.$$

Thus, a *nonlocal cosymmetry* in τ is a vector function $\tilde{\psi} = (\psi, \psi_X, \psi_Y)$ that satisfies the system:

$$\tilde{\ell}_{\mathcal{E}}^{*}(\psi) = \tilde{\ell}_X^{*}(\psi_X) + \tilde{\ell}_Y^{*}(\psi_Y),$$

$$\tilde{D}_x(\psi_X) + \tilde{D}_y(\psi_Y) = \frac{\partial Y}{\partial w}\psi_X + \frac{\partial X}{\partial w}\psi_Y.$$

In the case $\psi_X = \psi_Y = 0$, the function ψ satisfying the equation $\tilde{\ell}^*_{\mathscr{E}}(\psi) = 0$ is said to be a (*nonlocal*) *shadow* of cosymmetry.

Example 4.4 Consider the system:

$$v_y + wv_x = \frac{1}{w - v}, \qquad w_y + vw_x = \frac{1}{v - w}$$

(see Example 3.7) that is covered by the Gibbons-Tsarev equation. We consider it as an evolutionary equation (y being a "virtual time") and choose internal coordinates in the standard way. The linearization is

$$\ell_{\mathscr{E}} = \begin{pmatrix} D_y + wD_x - \dfrac{1}{(v-w)^2} & v_x + \dfrac{1}{(v-w)^2} \\ w_x + \dfrac{1}{(v-w)^2} & D_y + vD_x - \dfrac{1}{(v-w)^2} \end{pmatrix},$$

while

$$\ell^*_{\mathscr{E}} = \begin{pmatrix} -D_y - wD_x - w_x - \dfrac{1}{(v-w)^2} & w_x + \dfrac{1}{(v-w)^2} \\ v_x + \dfrac{1}{(v-w)^2} & -D_y - vD_x - v_x - \dfrac{1}{(v-w)^2} \end{pmatrix}.$$

Solving equation $\ell^*_{\mathscr{E}}(\xi) = 0$, we obtain the following generators $\xi_i = (\xi_i^v, \xi_i^w)$ in the space $\operatorname{cosym}(\mathscr{E})$:

$$\xi_0^v = 1, \qquad \xi_0^w = 1;$$

$$\xi_1^v = w + 2v, \qquad \xi_1^w = v + 2w;$$

$$\xi_2^v = 2y + w^2 + 2wv + 3v^2, \qquad \xi_2^w = 2y + v^2 + 2wv + 3w^2;$$

$$\begin{aligned} \xi_3^v &= x + 3yw + w^3 + 2(3y + w^2)v \\ &\quad + 3wv^2 + 4v^3, \end{aligned} \qquad \begin{aligned} \xi_3^w &= x + 3yv + v^3 + 2(3y + v^2)w \\ &\quad + 3vw^2 + 4w^3; \end{aligned}$$

$$\begin{aligned} \xi_4^v &= 4y^2 + 2xw + 4yw^2 + w^4 \\ &\quad + 2(2x + 4yw + w^3)v \\ &\quad + 3(w^2 + 4y)v^2 + 4wv^3 + 5v^4, \end{aligned} \qquad \begin{aligned} \xi_4^w &= 4y^2 + 2xv + 4yv^2 + v^4 \\ &\quad + 2(2x + 4yv + v^3)w \\ &\quad + 3(4y + v^2)w^2 + 4w^3v + 5w^4 \end{aligned}$$

and

$$\xi^v_{-5} = v_{xx}(v - w) + \frac{1}{2}(v_x - w_x)^2, \qquad \xi^w_{-5} = w_{xx}(w - v) + \frac{1}{2}(w_x - v_x)^2.$$

Let us denote the system under consideration by $\mathcal{E}$, the Gibbons-Tsarev equation by $\tilde{\mathcal{E}}$, $\tau\colon\tilde{\mathcal{E}} \to \mathcal{E}$ being the covering, and compare these results with those of Example 4.2. Taking into account Remark 4.1, one can see that passing from $\mathcal{E}$ to $\tilde{\mathcal{E}}$, the cosymmetry ξ_0 disappears, cosymmetries $\xi_1,\dots,\xi_4$ and ξ_{-5} transform to $\psi_0,\dots,\psi_{-5}$, respectively, while the cosymmetries ψ_0 and ψ_5 of the Gibbons-Tsarev equation are "purely nonlocal" in our covering.

4.1.1 Generating Functions of Conservation Laws

Let ω be a conservation law of $\mathcal{E}$. Then under the regularity condition formulated on p. 6, its generating function ψ_ω (see Sect. 3.1.1) is a cosymmetry of $\mathcal{E}$.

Remark 4.2 The converse statement is not valid generically, i.e., not every cosymmetry is a generating function of some conservation law. But one can state the following: if an equation is regular and normal (see p. 7 for the definition of normality) and the term $E_1^{2,n-1}$ of the *Vinogradov spectral sequence* (the $\mathcal{C}$-*spectral sequence*) vanishes, then the correspondence between conservation laws and cosymmetries is an isomorphism (see [18, 138, 139]). Examples can be found, e.g., in [3, 156] (see also [64–66]).

The explicit relation between ω and ψ_ω depends on a choice of internal coordinates in $\mathcal{E}$. In the simplest case, when $\mathcal{E}$

$$\frac{\partial u^j}{\partial t} = f^j\left(x, u, \dots, \frac{\partial^k u^\alpha}{\partial x^k}, \dots\right), \qquad j = 1, \dots, m,$$

is an evolutionary equation and internal coordinates u^j_k are chosen as it was indicated in Chap. 2, we have

$$\psi_\omega \equiv \delta(\omega) = \left(\frac{\delta X}{\delta u^1}, \dots, \frac{\delta X}{\delta u^m}\right)$$

for a conservation law $\omega = X\,dx + T\,dt$. Here

$$\frac{\delta X}{\delta u^j} = \sum_k (-1)^k D_x^k\left(\frac{\partial X}{\partial u^j_k}\right)$$

denotes the variational derivative and δ is the Euler operator on $\mathcal{E}$.

Example 4.5 Consider the conservation law

$$\omega = u\,dt + \left(\frac{1}{2}u^2 + u_x\right)\,dt$$

of the Burgers equation. Its generating function is $\psi_\omega = \delta(u) = \partial u/\partial u = 1$.

Remark 4.3 A scalar equation $\mathscr{E} = \{F = 0\} \subset J^\infty(n, 1)$ admits the cosymmetry $\psi = 1$ if and only if the equation itself is in the *conserved form*, i.e.,

$$F\,dx^1 \wedge \cdots \wedge dx^n = d_h(\omega)$$

for some $\omega \in \Lambda_h^{n-1}(J^\infty(n, 1))$.

More generally, let $\mathscr{E} = \{F^1 = \cdots = F^r = 0\} \subset J^\infty(n, m)$ admit a cosymmetry $\alpha = (\alpha^1, \ldots, \alpha^r)$, $\alpha^j \in \mathbb{R}$. Then $\sum_j \alpha^j F^j\,dx^1 \wedge \cdots \wedge dx^n = d_h(\omega)$ for some differential $(n-1)$-form on $J^\infty(n, m)$. This is a particular case of the general definition (4.3) below.

Example 4.6 Consider the conservation laws $\omega_i = P_i\,dx + Q_i\,dy$ of the Gibbons-Tsarev equation (Example 3.2). Then the corresponding generating functions (see Example 4.2) are computed by $\psi_{\omega_i} = \partial P_i/\partial z_y$.

Remark 4.4 Actually, the Euler operator is a particular coordinate form of the differential $\delta\colon E_1^{0,n-1} \to E_1^{1,n-1}$ in the $\mathscr{C}$-spectral sequence for jet spaces.

Example 4.7 The two-component version of the Gibbons-Tsarev equation (see Example 4.4) possesses two obvious conservation laws: the first one, $\omega_0 = (u + v)\,dx - uv\,dy$, was already mentioned, and its generating function is $\xi_0 = (1, 1)$. The second,

$$\omega_1 = (v^2 + vw + w^2)\,dx - (x + v^2w + w^2v)\,dy,$$

can be easily obtained multiplying the first equation by v and the second one by w and adding the results. The generating function of ω_1 is $\xi_1 = (2v + w, 2w + v)$.

4.1.2 *Reconstruction of Conservation Laws by Their Generating Functions*

Integrating by parts Eq. (3.2), we can reduce the order of Δ to zero and thus obtain the relation:

$$D_1(a_1) - D_2(a_2) + \cdots + (-1)^{n-1}D_n(a_n) = \psi^1F^1 + \cdots + \psi^rF^r, \qquad (4.3)$$

where $\psi^1, \ldots, \psi^r$ are functions on the space of jets. Thus, to find conservation laws corresponding to a cosymmetry $\psi = (\psi^1, \ldots, \psi^r)$, one needs to solve Eq. (4.3) with respect to the functions a_i. In the evolutionary case, this equation is solvable if and only if the operator $\ell_\psi : \mathscr{F}(n, m; r) \to \mathscr{F}(n, m; r)$ is self-adjoint. Of course, all the solutions, if any, differ by a trivial conservation law.

Example 4.8 Consider the cosymmetry $\psi_1(T) = T/u_y^2$ of the universal hierarchy equation (Example 4.3). Then, by (4.3), one has

$$\frac{T}{u_y^2}(u_{yy} - u_t u_{xy} + u_y u_{xt}) = -\left(\frac{T}{u_y}\right)_y - \left(\frac{Tu_t}{u_y}\right)_x$$

Thus, the form

$$\omega_T = \frac{T}{u_y}\,dx \wedge dt - \frac{Tu_t}{u_y}\,dy \wedge dt$$

is a two-component conservation law of the universal hierarchy equation for any smooth function $T = T(t)$.

Remark 4.5 Let $\mathscr{E}$ be an equation in independent variables $x^1, \dots, x^n$, $n > 2$ and

$$\omega = X_1\,dx^1 \wedge dx^3 \wedge \cdots \wedge dx^n + X_2\,dx^2 \wedge dx^3 \wedge \cdots \wedge dx^n$$

be a two-component conservation law of the equation $\mathscr{E}$. Then, evidently, any form $\Phi\omega$, where $\Phi = \Phi(x^2, \dots, x^n)$, is a conservation law as well.

Example 4.9 Let us find the conservation law that corresponds to the cosymmetry

$$\xi_2 = (2y + w^2 + 2wv + 3v^2, 2y + v^2 + 2wv + 3w^2)$$

in Example 4.4. Using Eq. (4.3), we get

$$\begin{aligned}(2y + w^2 + 2wv + 3v^2)\left(v_y + wv_x + \frac{1}{v - w}\right)&\\ + (2y + v^2 + 2wv + 3w^2)\left(w_y + vw_x - \frac{1}{v - w}\right)&\\ = \left(\frac{2}{3}(v^3 + w^3) + 2y(v + w) + \frac{2}{3}(v + w)^3\right)_y + \left(2yvw + vw(v^2 + w^2) + (vw)^2\right)_x&\end{aligned}$$

Hence, the form

$$\begin{aligned}\omega_2 = \left(\frac{2}{3}(v^3 + w^3) + 2y(v + w) + \frac{2}{3}(v + w)^3\right)dx&\\ - \left(2yvw + vw(v^2 + w^2) + (vw)^2\right)dy&\end{aligned}$$

is the desired conservation law.

Example 4.10 Consider the cosymmetry $\psi_{-5} = u_{xxx}$ of the Gibbons-Tsarev equation (Example 4.2). One has

$$u_{xxx}(u_{yy} + u_x u_{xy} - u_y u_{xx} + 1)$$
$$= D_y\left(u_y u_{xxx} - \frac{1}{2}u_x u_{xx}^2\right) - D_x\left(u_y u_{xxy} + \frac{1}{2}u_y u_{xx}^2 - u_x u_{xx} u_{xy} - \frac{1}{2}u_{xy}^2 - u_{xx}\right),$$

and thus ψ_{-5} is the generating function of the conservation law $\omega_{-5} = P_{-5}\,dx + Q_{-5}\,dy$ from Example 3.2.

4.2 Examples

We deal here with two different problems: starting from conservation laws, we would like to find the corresponding generating functions, and starting from the equation, we would like to first compute the generating functions of conservation laws and then the corresponding conservation laws.

4.2.1 Korteweg-de Vries Equation

We look for cosymmetries as solutions of the equation:

$$D_t(\psi) = uD_x(\psi) + D_x^3(\psi),$$

see Sect. 2.3.1. The program file is kdv_csy1.red. After defining the adjoint linearization

```
mk_cdiffop(ellstar_k,1,{1},1);
for all csy let ellstar_k(csy) =
  td(csy,t) - u*td(csy,x) - td(csy,x,3);
```

we require that cosymmetries are functions of dependent variables only:

```
for each el in dep_var do depend psi,el;
```

The remaining variables are collected in the following list:

```
split_vars:=for i:=1:total_order join
  selectvars(0,i,dep_var,all_parametric_der);
```

The equation is generated as follows:

```
system_eq:=ellstar_k(psi);
```

Then, we run CRACK on the above system:

```
unk:={psi};
load_package crack;
crack_results:=crack(system_eq,{},unk,split_vars);
```

The only solutions, as it is well known, are

```
psi=c_1 + c_2*u
```

where c_1 and c_2 are two constants.

4.2.2 Dispersionless Boussinesq System

Recall that cosymmetries of the dispersionless Boussinesq equation are solutions to the system:

$$\begin{pmatrix} D_t & 3wD_x + 2w_1 & -D_x \\ -D_x & D_t & 0 \\ -wD_x & uD_x - 2u_1 & D_t \end{pmatrix} \begin{pmatrix} \psi^1 \\ \psi^2 \\ \psi^3 \end{pmatrix} = \begin{pmatrix} 0 \\ 0 \\ 0 \end{pmatrix} ;$$

see Sect. 2.3.2.

If ω_1, ω_2, and ω_3 are the conservation laws presented in Sect. 3.2.2, then

$$\psi_{\omega_1} = (0,0,1), \quad \psi_{\omega_2} = (1,0,0), \quad \psi_{\omega_3} = (0,1,2w).$$

After loading the Boussinesq equation, we can easily find the above generating functions by computing the variational derivatives of the densities of the conservation laws in Sect. 3.2.2. The CDE implementation is in the program bou_csy1.red:

```
pvar_df(0,w,u);
0
pvar_df(0,w,v);
0
pvar_df(0,w,w);
1
```

and so on. Note the use of the function pvar_df, which just computes the variational derivative of the second argument with respect to an (even or odd) dependent variable at the third argument. The first argument can be 0 or 1 depending on the parity of the variable with respect to which the variational derivative is computed.

If we wish to verify that these quantities are in the kernel of the adjoint linearization, just load them by

```
mk_superfun(lbou_sf,1,3);
mk_superfun(lbou_star_sf,1,3);
```

```
in "bou_ell1_res.red";
conv_superfun2cdiff(lbou_star_sf,lbou_star);
```

and apply the above matrix operator to a generating function as follows:

```
cldens_1:=w;
cldens_2:=u;
cldens_3:=v+w**2;

genfun_1:=for each el in dep_var collect
  pvar_df(0,cldens_1,el);
genfun_2:=for each el in dep_var collect
  pvar_df(0,cldens_2,el);
genfun_3:=for each el in dep_var collect
  pvar_df(0,cldens_3,el);

res_1:=for i:=1:ncomp collect
  for j:=1:ncomp sum
    lbou_star(i,j,part(genfun_1,j));

res_2:=for i:=1:ncomp collect
  for j:=1:ncomp sum
    lbou_star(i,j,part(genfun_2,j));

res_3:=for i:=1:ncomp collect
  for j:=1:ncomp sum
    lbou_star(i,j,part(genfun_3,j));
```

The result will be three lists of three zeros each.

4.2.3 Camassa-Holm Equation

Recall (cf. Sect. 2.3.3) that cosymmetries of the Camassa-Holm equation are solutions of

$$\alpha(D_t(\psi) + 3uD_x(\psi)) - D_tD_x^2(\psi) = u_{xx}D_x(\psi) + u_xD_x^2(\psi) + uD_x^3(\psi),$$

where $\alpha = \text{const} \in \mathbb{R}$.

Since this example is non-evolutionary, we cannot use the above approach to construct cosymmetries from conservation laws. However, we can find cosymmetries as functions which lie in the kernel of the adjoint linearization. The CDE implementation is in the program ch_csy1.red.

For the Camassa-Holm equation, we can construct the jet space as before and use the following weights for the variables:

```
deg_indep_var:={-2,-1};
deg_dep_var:={1};
deg_odd_var:={1};
```

Note that with the above weights, we have to assume that `alpha` has weight 2 for consistency. Then we must load the adjoint linearization:

```
mk_superfun(lch_star_sf,1,1);
lch_star_sf(1):= - alpha*p_t - 3*alpha*p_x*u
  + p_2x*u_x + p_3x*u + p_t2x + p_x*u_2x;
conv_superfun2cdiff(lch_star_sf,lch_star);
```

and generate an ansatz for a cosymmetry `psi` with the usual mechanism. Here we assume that the variable `ansatz` contains weighted monomials of degrees ranging from 0 to 6. Note that in order to generate a correct ansatz, we need to include `alpha` within the graded variables. Here we face a choice: we might generate graded monomials using `alpha` in the list `l_grad_der`, or we can include `alpha` in the list of graded monomials in the appropriate place. We make the second choice as we do not want to consider solutions which are just multiple of `alpha`, even if we will exclude some solutions in this way. More precisely, we generate the ansatz in the following way:

```
l_grad_der:=der_deg_ordering(0,all_parametric_der);
% List of monomials of weight <= 6
gradmon:=graded_mon(1,6,l_graded_der);
gradmon:=part(gradmon,2):=alpha . part(gradmon,2);
gradmon:={1} . gradmon;
ansatz:=for each el in gradmon join el;
phi:=for each el in ansatz sum
  c(ctel:=ctel+1)*el;
```

After setting up the equation

```
equ 1:=num lch_star(1,1,phi);
```

and solving it in the standard way, we get

```
phi;

 (15*c(6)*alpha*u**3 - 12*c(6)*u**2*u_2x
  - 6*c(5)*u*u_tx - 3*c(6)*u*u_x**2 + 6*c(6)*u_2t
  + 6*c(6)*u_t*u_x + 15*c(4)*alpha*u**2
  - 10*c(4)*u*u_2x - 10*c(4)*u_tx - 5*c(4)*u_x**2
  + 15*c(3)*alpha**2 + 15*c(2)*alpha*u +
  15*c(1)*alpha)/(15*alpha);
```

4.2.4 Gibbons-Tsarev Equation

The program gt_lcl1.red of Sect. 3.2.4 for finding conservation laws of the Gibbons-Tsarev equation can be easily modified into the program gt_csy1.red for finding cosymmetries of the same equation. The initialization and the monomials generated for the ansatz are exactly the same as those of gt_lcl1.red. Then we generate an ansatz for a cosymmetry which is a polynomial of degree at most three in independent variables, dependent variables, and their first-order derivatives:

```
psi_list:=for each el in list_mon collect
   (c(ctel:=ctel+1)*el);
psi:=(part(psi_list,0):=plus);
```

In order to speed up the computation (this holds especially if one considers higher polynomial degrees for the ansatz), it is better to pre-calculate the total x- and y-derivatives:

```
psi_x_list:=for each el in psi_list collect td(el,x);
psi_y_list:=for each el in psi_list collect td(el,y);
psi_x:=(part(psi_x_list,0):=plus);
psi_y:=(part(psi_y_list,0):=plus);
```

Then we can formulate the equation:

```
equ 1:=td(psi_y,y) + u_x*td(psi_y,x)
   - u_y*td(psi_x,x) - 2*u_xy*psi_x + 2*u_2x*psi_y;
```

which can be solved in the usual way. We get

```
(4*c(6)*u_x**3 + 6*c(6)*u_x*u_y + 8*c(6)*u_x*y
  + 2*c(6)*x + 2*c(3)*u_x + 3*c(2)*u_x**2
  + 2*c(2)*u_y + 3*c(2)*y + 2*c(1))/2;
```

which are exactly the first cosymmetries of Example 4.6.

The problem of reconstructing the conservation laws from the above cosymmetries is more difficult than its inverse problem (i.e., finding the generating function for a conservation law), especially in the case of non-evolution equations. It is clear that we could compute the variational derivatives of the densities of the conservation laws in Sect. 3.2.4 and compare them with the above cosymmetries. However, let us proceed backward from cosymmetries to conservation laws by a direct integration of (1.19). We describe the program file gt_csy2.red.

First of all, we stress that (1.19) must be solved on the ambient jet space. This means that CDE has to be initialized as follows:

```
indep_var:={x,y};
dep_var:={u};
total_order:=10;

principal_der:={u_2y};
```

```
de:={ - u_x*u_xy + u_y*u_2x - 1};
```

```
cde({indep_var,dep_var,{},total_order},{});
```

Then, we load the above cosymmetries and consider only the one that corresponds to `c(3)`:

```
psi:=df(psi,c(3));
```

We look for polynomial conservation laws with respect to the independent variables, dependent variables, and their first-order derivatives. The form is `cy dx + cx dy`. We generate the list of monomials of algebraic degree 4, as the right-hand side of (1.19) has degree 4 in our case.

```
top_degree:=4;
even_vars:=for i:=0:1 join
  selectvars(0,i,dep_var,all_parametric_der);
all_vars:=append(indep_var,even_vars);
list_mon:=1 . mkallmon(top_degree,all_vars);
```

Then we generate the ansatz for `cy` and `cx` and precompute their total derivatives:

```
cx_list:=for each el in list_mon collect
   (c(ctel:=ctel+1)*el);
cy_list:=for each el in list_mon collect
   (c(ctel:=ctel+1)*el);
cx_x_list:=for each el in cx_list collect td(el,x);
cy_y_list:=for each el in cy_list collect td(el,y);
cx:=(part(cx_list,0):=plus);
cy:=(part(cy_list,0):=plus);
cx_x:=(part(cx_x_list,0):=plus);
cy_y:=(part(cy_y_list,0):=plus);
```

The Gibbons-Tsarev equation is

```
gt_eq:=first(principal_der) - first(de);
```

and the equation for conservation laws is

```
equ 1:=cy_y - cx_x - psi*gt_eq;
```

After the usual integration procedure, we find that `cy` and `cx` are defined by a "fixed" part and a part that depends on constants which is evidently made by trivial conservation laws:

```
cy := (12*c(183)*x**4 + 12*c(148)*x**3 + 12*c(133)
*x**2 + 12*c(132)*x + 12*c(127) + 4*c(101)*u**3 +
12*c(101)*u**2*u_x*x + 12*c(99)*u**3*u_x + 12*c(96
)*u**2*x + 12*c(96)*u*u_x*x**2 + 36*c(92)*u*x**2 +
 12*c(92)*u_x*x**3 + 3*c(61)*y**4 + 8*c(59)*x*y**3
 + 18*c(58)*x**2*y**2 + 48*c(57)*x**3*y + 6*c(47)*
```

```
u**2 + 12*c(47)*u*u_x*x + 12*c(45)*u**2*u_x + 24*c
(42)*u*x + 12*c(42)*u_x*x**2 + 12*c(31)*u**2*y +
24*c(31)*u*u_x*x*y + 12*c(30)*u*u_x*y**2 + 36*c(29
)*u**2*u_x*y + 6*c(28)*u*y**2 + 6*c(28)*u_x*x*y**2
 + 4*c(27)*u_x*y**3 + 24*c(26)*u*x*y + 12*c(26)*
u_x*x**2*y + 4*c(25)*y**3 + 12*c(23)*x*y**2 + 36*c
(22)*x**2*y + 12*c(21)*u + 12*c(21)*u_x*x + 12*c(
19)*u*u_x + 12*c(12)*u*y + 12*c(12)*u_x*x*y + 6*c(
11)*u_x*y**2 + 24*c(10)*u*u_x*y + 6*c(9)*y**2 + 24
*c(7)*x*y + 12*c(6)*y + 12*c(4)*u_x*y + 12*c(2)*
u_x + 6*u_x**3 + 12*u_x*u_y + 12*u_x*y)/12;

cx := (6*c(101)*u**2*u_y*x + 6*c(99)*u**3*u_y + 6*
c(96)*u*u_y*x**2 + 6*c(92)*u_y*x**3 + 6*c(61)*x*y
**3 + 6*c(60)*y**4 + 6*c(59)*x**2*y**2 + 6*c(58)*x
**3*y + 6*c(57)*x**4 + 6*c(47)*u*u_y*x + 6*c(45)*u
**2*u_y + 6*c(42)*u_y*x**2 + 6*c(31)*u**2*x + 12*c
(31)*u*u_y*x*y + 6*c(30)*u**2*y + 6*c(30)*u*u_y*y
**2 + 6*c(29)*u**3 + 18*c(29)*u**2*u_y*y + 6*c(28)
*u*x*y + 3*c(28)*u_y*x*y**2 + 6*c(27)*u*y**2 + 2*c
(27)*u_y*y**3 + 6*c(26)*u*x**2 + 6*c(26)*u_y*x**2*
y + 6*c(25)*x*y**2 + 6*c(24)*y**3 + 6*c(23)*x**2*y
 + 6*c(22)*x**3 + 6*c(21)*u_y*x + 6*c(19)*u*u_y +
6*c(12)*u*x + 6*c(12)*u_y*x*y + 6*c(11)*u*y + 3*c(
11)*u_y*y**2 + 6*c(10)*u**2 + 12*c(10)*u*u_y*y + 6
*c(9)*x*y + 6*c(8)*y**2 + 6*c(7)*x**2 + 6*c(6)*x +
 6*c(5)*y + 6*c(4)*u + 6*c(4)*u_y*y + 6*c(2)*u_y +
 6*c(1) + 3*u_x**2*u_y + 3*u_y**2 + 6*u_y*y)/6;
```

It is quite easy to realize that cy and cx correspond to P_1 and Q_1 of Example 3.2. Moreover, it is not difficult to extend the same procedure to other cosymmetries. This example also shows how fast the complexity of symbolic computations in this subject can increase!

4.2.5 Multi-dimensional Examples

Let us pass to examples in dimensions higher than two.

4.2.5.1 Universal hierarchy Equation

We describe the program uh_csy1.red and initialize CDE as in the Sect. 3.2.5.1. The adjoint linearization of the UHE is defined by

```
mk_cdiffop(ellstar_uh,1,{1},1);
for all psi let
  ellstar_uh(1,1,psi) = td(psi,y,2) - u_t*td(psi,x,y)
    - u_y*td(psi,x,t) - 2*u_tx*td(psi,y)
    + 2*u_xy*td(psi,t);
```

see Example 2.18.

Then, we define an ansatz for psi as a function of independent variables, dependent variables, and their first derivatives:

```
ansatz_vars:=for i:=0:1 join
  selectvars(0,i,dep_var,all_parametric_der);
for each el in ansatz_vars do depend psi,el;
depend psi,t,x,y;
```

Let us introduce the list of the variables with respect to which the equation is polynomial:

```
split_vars:=
  cde_difflist(all_parametric_der,ansatz_vars);
```

The command cde_difflist computes the list of elements which are members of the first list and are not members of the second list. Both lists must be sets, i.e., lists of mutually different elements. The unknown function is

```
unk:={psi};
```

and the equation $\ell^*_{\mathcal{E}}(\psi) = 0$ translates into

```
lin_eq:={ellstar_uh(1,1,psi)};
```

Collecting the coefficients of the polynomial, we obtain the following system:

```
system_eq:=splitvars_list(lin_eq,split_vars);
```

The command splitvars_list splits a system of equations lin_eq which are polynomial with respect to the variables split_vars into the list of all coefficients. We pass the system to the PDE solver CRACK:

```
load_package crack;
crack_sol:=crack(system_eq,{},unk,{});
```

We obtain

```
crack_sol := {{{},
  {psi=(3*df(c_11,x)*u*u_y**3 - df(c_25,t)*u_y*y +
    6*c_11*u_x*u_y**3 + 2*c_25*u_t + 6*c_26*u_y**3
    - 6*c_31*u_y + 3*c_33*u_y**3)/(6*u_y**3)},
  {c_11, c_25, c_26, c_31, c_33},
  {}}};
```

We should only check the dependency of the integration quantities on the variables. We use the following command:

```
fargs(c_11);

{x};
```

(the command is provided by the package CRACK). The quantities `c_11`, `c_26`, and `c_33` depend on x, while the quantities `c_25` and `c_31` depend on t.

4.2.5.2 Khokhlov-Zabolotskaya Equation

Recall that (potential) Khokhlov-Zabolotskaya (or dispersionless Kadomtsev-Petviashvili) equation (see Sect. 3.2.5.2) is

$$v_{yy} = -v_{xx}v_x + v_{tx},$$

or, in the evolutionary form,

$$v_y = w, \qquad w_y = -v_{xx}v_x + v_{tx}.$$

The linearization of the obtained system is

$$\ell_{\mathcal{E}} = \begin{pmatrix} D_y & -1 \\ v_{xx}D_x + v_xD_x^2 - D_xD_t & D_y \end{pmatrix},$$

while its adjoint is of the form

$$\ell_{\mathcal{E}}^* = \begin{pmatrix} -D_y & v_{xx}D_x + v_xD_x^2 - D_xD_t \\ -1 & -D_y \end{pmatrix}.$$

We can load the results from the program file pkz_lcl1.red described in Sect. 3.2.5 into the new program file pkz_csy1.red. We removed trivial conservation laws from those obtained in the computation; however, the remaining conservation laws could be linearly dependent up to trivial conservation laws. A simple way to know how many independent conservation laws are contained in the solution space that we obtained so far is computing its generating functions. We can do this by the command:

```
genfun:=
  {pvar_df(0,cynontriv,v),pvar_df(0,cynontriv,w)};
```

Then, we extract all integration constants in the conservation laws:

```
c_list:=cde_mkset(
      append(cdiff_get_kernels(cxnontriv,c),
      cdiff_get_kernels(cynontriv,c))
);
```

Here we use the CDIFF function cdiff_get_kernels in order to extract all instances of the operator c in the lists cxnontriv and cynontriv and the CDE function cde_mkset in order to make a set out of the result. A set is a list with no duplicate items.

Finally, we make a list of pairs of integration constants and related generating functions:

```
genfun_el:=for each el in c_list collect
  {el,for each ell in genfun collect df(ell,el)};
```

By direct inspection of the list genfun_el, we see that there are only two independent generating functions: $(0, 1)$, corresponding to ω_1 and the vector function whose components are

```
1: pvar_df(0,df(cynontriv,c(681)),v);

 - 2*t*v_2x*v_x + 2*t*v_tx + v_x + w_x*y

2: pvar_df(0,df(cynontriv,c(681)),w);

- 2*t*w - v_x*y
```

corresponding to ω_2.

Chapter 5
Symmetries

Abstract Symmetries of $\mathscr{E} \subset J^\infty(n, m)$ are vector fields that preserve solutions of $\mathscr{E}$. Effectively, this means that they preserve the Cartan distribution on the equation at hand. We discuss symmetries and related notions in this chapter and describe solutions to Problems 1.5, 1.13, and 1.14.

5.1 General Theory

We expose the theory in two versions: local and nonlocal.

5.1.1 Local Symmetries

A vertical vector field

$$X = \sum_{\sigma,j\in\mathbb{I}} b^j_\sigma \frac{\partial}{\partial u^j_\sigma}$$

where $\mathbb{I}$, as before, denotes the chosen set of all internal coordinates in $\mathscr{E}$, is a *symmetry* if

$$[X, D_i] = 0, \qquad i = 1, \dots, n.$$

When the equation under consideration satisfies the generality condition (see p. 6), any symmetry is of the form

$$\mathbf{E}_\varphi = \sum_{\sigma,j\in\mathbb{I}} D_\sigma(\varphi^j)\frac{\partial}{\partial u^j_\sigma}, \tag{5.1}$$

J. Krasil'shchik et al., *The Symbolic Computation of Integrability Structures for Partial Differential Equations*, Texts and Monographs in Symbolic Computation,
https://doi.org/10.1007/978-3-319-71655-8_5

where the vector function $\varphi = (\varphi^1, \dots, \varphi^m) \in \mathscr{F}(\mathscr{E}; m)$ is called the *generating function* of the symmetry and satisfies the linearized equation

$$\ell_{\mathscr{E}}(\varphi) = 0. \tag{5.2}$$

Fields of the form $\mathbf{E}_\varphi$ are called *evolutionary vector fields*. These fields are identified with their generating functions.

Example 5.1 As a first example, let us consider the Burgers equation $u_t = uu_x + u_{xx}$. Its symmetries are solutions to the equation

$$D_t(\varphi) = u_x\varphi + uD_x(\varphi) + D_x^2(\varphi)$$

where $\varphi = \varphi(x, t, u, u_1, \dots, u_k)$. Let us say that k is the *order* of the symmetry φ. Solving this equation for small values of k, we obtain the following symmetries:

$$k = 1{:}\varphi_1^0 = u_1, \qquad \varphi_1^1 = 1 + tu_1;$$

$$k = 2{:}\varphi_2^0 = uu_1 + u_2, \qquad \varphi_2^1 = \frac{1}{2}u + \left(\frac{1}{2}x + tu\right)u_1 + tu_2,$$

$$\varphi_2^2 = x + tu + (xt + ut^2)u_1 + t^2u_2.$$

Here φ_1^0 is the *x-translation*, φ_2^0 is the *t-translation*, φ_1^1 is the *Galilean boost*, while φ_2^1 is the *scaling symmetry*.

Further computations reveal existence of four symmetries of order 3:

$$\begin{aligned}
\varphi_3^0 &= \frac{3}{4}u^2u_1 + \frac{3}{2}u_1^2 + \frac{3}{2}uu_2 + u_3,\\
\varphi_3^1 &= \frac{1}{4}u^2 + \frac{1}{4}(2xu + 3u^2t)u_1 + \frac{3}{2}tu_1^2 + \frac{1}{2}(x + 3tu)u_2 + tu_3,\\
\varphi_3^2 &= \frac{1}{2}xu + \frac{1}{2}u^2t + \frac{1}{4}(x^2 + 6t + 4txu + 3t^2u^2)u_1 + \frac{3}{2}t^2u_1^2 + \frac{1}{2}t(2x + 3tu)u_2 + t^2u_3,\\
\varphi_3^3 &= \frac{3}{4}(2t + t^2u^2 + x^2) + \frac{3}{4}t(4t + t^2u^2 + 2txu + x^2)u_1 + \frac{3}{2}t^3u_1^2 + \frac{3}{2}t^2(tu + x)u_2\\
&\quad + \frac{3}{2}tux + t^3u_3.
\end{aligned}$$

As we shall see below (Remark 5.3), existence of symmetries φ_2^2 and φ_3^1 implies existence of infinite number of symmetries for the Burgers equation.

Remark 5.1 Due to existence of the scaling symmetry, one can assign the following weights to the jet variables:

$$[x] = 1, \qquad [t] = 2, \qquad [u_k] = -(k + 1),$$

and define the weight of a monomial as the sum of its factors weights. Then the weights of the computed symmetries[1] are as follows:

$$[\varphi_1^0] = -1,\ [\varphi_1^1] = 1,$$
$$[\varphi_2^0] = -2,\ [\varphi_2^1] = 0,\quad [\varphi_2^2] = 2,$$
$$[\varphi_3^0] = -3,\ [\varphi_3^1] = -1,\ [\varphi_3^2] = 1,\ [\varphi_3^3] = 3.$$

Example 5.2 The Gibbons-Tsarev equation $u_{yy} + u_x u_{xy} - u_y u_{xx} + 1 = 0$ has the linearization

$$\ell_{\mathcal{E}} = D_y^2 + u_x D_x D_y - u_y D_x^2 + u_{xy} D_x - u_{xx} D_y$$

and admits the following five symmetries:

$\varphi_{-4} = 1,$	u-translation,
$\varphi_{-3} = u_x,$	x-translation,
$\varphi_{-2} = u_y,$	y-translation,
$\varphi_{-1} = yu_x - 2x,$	generalized Galilean boost,
$\varphi_0 = 3xu_x + 2yu_y - 4u,$	scaling.

Again, existence of the scaling symmetry φ_0 allows one to introduce the weights

$$[x] = 3, \qquad [y] = 2, \qquad [u] = 4,$$

and one has $[\varphi_i] = i$, $i = -4, \dots, 0$.

Example 5.3 The linearization of the universal hierarchy equation $u_{yy} = u_t u_{xy} - u_y u_{tx}$ is

$$\ell_{\mathcal{E}} = D_y^2 - u_t D_x D_y + u_y D_x D_t - u_{xy} D_t + u_{xt} D_y,$$

and the equation possesses the symmetries

$$\theta_0(X) = Xu_x - X'u, \qquad \theta_1(X) = X,$$
$$\varphi_0(T) = Tu_t + T'yu_y, \qquad \varphi_{-1}(T) = Tu_y,$$
$$\upsilon_0 = yu_y + u,$$

where $X = X(x)$ and $T = T(t)$ are arbitrary smooth functions and "prime" stands for the corresponding derivative. In particular,

[1] As *vector fields* not as *functions*!

$$\theta_0(1) = u_x, \qquad \varphi_0(1) = u_t, \qquad \varphi_{-1}(1) = u_y, \qquad \theta_1(1) = 1$$

are x-, y-, t-, and u-translations, respectively, and

$$\theta_0(x) = xu_x - u, \qquad \varphi_0(t) = tu_t + yu_y, \qquad \upsilon_0 = yu_y + u$$

are three independent scaling symmetries. Consequently, the equation admits three systems of weights

$$\begin{aligned}
&[x]^{(1)} = 1 \ [y]^{(1)} = 0 \ [t]^{(1)} = 0 \ [u]^{(1)} = 1;\\
&[x]^{(2)} = 0 \ [y]^{(2)} = 1 \ [t]^{(2)} = 0 \ [u]^{(2)} = -1;\\
&[x]^{(3)} = 0 \ [y]^{(3)} = 0 \ [t]^{(3)} = 1 \ [u]^{(3)} = 1.
\end{aligned}$$

In what follows, the second system of weights,

$$[\theta_0(X)]^{(2)} = 0, \quad [\theta_1(X)]^{(2)} = 1,$$
$$[\varphi_0(T)]^{(2)} = 0, \quad [\varphi_{-1}(T)]^{(2)} = -1, \quad [\upsilon_0]^{(2)} = 0,$$

will be used.

Example 5.4 Let us finally describe symmetries $\varphi = (V, W)$ of the two-component version of the Gibbons-Tsarev equation

$$v_y + wv_x = \frac{1}{w - v}, \qquad w_y + vw_x = \frac{1}{v - w}.$$

These symmetries are to satisfy the system

$$\begin{aligned}
D_y(V) + wD_x(V) + v_x W &= \frac{V - W}{(v - w)^2},\\
D_y(W) + vD_x(W) + w_x V &= \frac{W - V}{(v - w)^2}.
\end{aligned}$$

The space of solutions is spanned by the elements

$$\begin{aligned}
&\varphi_{-3} = (v_x, w_x), \qquad \varphi_{-2} = (v_y, w_y), \qquad \varphi_{-1} = (1 - yv_x, 1 - yw_x),\\
&\varphi_0 = (3xv_x + 2yv_y - v, 3xw_x + 2yw_y - w).
\end{aligned}$$

The first two symmetries are x- and y-translations, φ_{-1} is the Galilean boost, and φ_0 is the scaling symmetry. It allows one to introduce the weights

$$[x] = 3, \qquad [y] = 2, \qquad [v] = 1, \qquad [w] = 1,$$

and one has $[\varphi_i] = i$, for $i = -3, \ldots, 0$.

5.1.2 *Jacobi Bracket*

Symmetries of $\mathscr{E}$ form a Lie algebra with respect to commutator which is denoted by $\operatorname{sym}(\mathscr{E})$. The commutator of two symmetries $\mathbf{E}_{\varphi_1}$ and $\mathbf{E}_{\varphi_2}$ is a symmetry, and thus we have

$$[\mathbf{E}_{\varphi_1}, \mathbf{E}_{\varphi_2}] = \mathbf{E}_{\varphi}$$

for some uniquely defined φ. We call φ the *Jacobi bracket* of φ_1 and φ_2 and use the notation $\varphi = \{\varphi_1, \varphi_2\}$. The Jacobi bracket can be computed by the formula

$$\{\varphi_1, \varphi_2\} = \mathbf{E}_{\varphi_1}(\varphi_2) - \mathbf{E}_{\varphi_2}(\varphi_1)$$

and in local coordinates its components are

$$\{\varphi_1, \varphi_2\}^j = \sum_{\mathbb{I}} \left(D_\sigma(\varphi_1^\alpha) \frac{\partial \varphi_2^j}{\partial u_\sigma^\alpha} - D_\sigma(\varphi_2^\alpha) \frac{\partial \varphi_1^j}{\partial u_\sigma^\alpha} \right). \tag{5.3}$$

Another way to compute the bracket is

$$\{\varphi_1, \varphi_2\} = \ell_{\varphi_2}(\varphi_1) - \ell_{\varphi_1}(\varphi_2) = \left(\mathbf{E}_{\varphi_1} - \ell_{\varphi_1}\right)(\varphi_2).$$

Thus, the operator $L_\varphi = \mathbf{E}_\varphi - \ell_\varphi \colon \operatorname{sym}(\mathscr{E}) \to \operatorname{sym}(\mathscr{E})$ is the *Lie derivative* that determines the action of φ on $\operatorname{sym}(\mathscr{E})$.

Remark 5.2 The action on cosymmetries is defined in a more complicated way. Let $\mathscr{E} = \{F = 0\} \subset J^\infty(n, m)$, $F \in P$, and $\varphi \in \operatorname{sym}(\mathscr{E})$. Let $\bar{\varphi}$ be an extension of φ to $J^\infty(n, m)$, $\bar{\varphi}|_{\mathscr{E}} = \varphi$. Consider the evolutionary vector field $\mathbf{E}_{\bar{\varphi}}$ on $J^\infty(n, m)$. Then, by the regularity condition, one has

$$\mathbf{E}_{\bar{\varphi}}(F) = \bar{\Delta}(F), \qquad \bar{\Delta} \colon P \to P,$$

where $\bar{\Delta}$ is a $\mathscr{C}$-differential operator. Then

$$L_\varphi(\psi) = (\mathbf{E}_\varphi + \Delta^*)(\psi), \qquad \psi \in \operatorname{cosym}(\mathscr{E}),$$

where $\Delta = \bar{\Delta}\big|_{\mathscr{E}}$. In particular, when $\mathscr{E}$ is an evolution equation, we obtain

$$L_\varphi = \mathbf{E}_\varphi + \ell_\varphi^*.$$

Example 5.5 The Jacobi brackets of first- and second-order symmetries for the Burgers equation (see Example 5.1) are

$$\begin{aligned}\{\varphi_1^0,\varphi_1^1\}=0,\ \{\varphi_1^0,\varphi_2^0\}=0,\ \ &\{\varphi_1^0,\varphi_2^1\}=-\tfrac{1}{2}\varphi_1^0,\ \{\varphi_1^0,\varphi_2^2\}=-\varphi_1^1,\\ \{\varphi_1^1,\varphi_2^0\}=\varphi_1^1,\ &\{\varphi_1^1,\varphi_2^1\}=\tfrac{1}{2}\varphi_1^1,\ \ \{\varphi_1^1,\varphi_2^2\}=0,\\ &\{\varphi_2^0,\varphi_2^1\}=-\varphi_2^0,\ \ \{\varphi_2^0,\varphi_2^2\}=-2\varphi_2^1,\\ &\{\varphi_2^1,\varphi_2^2\}=-\varphi_2^2.\end{aligned}$$

Remark 5.3 So, we see that symmetries of order ≤ 2 form a Lie algebra with respect to the Jacobi bracket. But if we compute brackets for symmetries of higher order, we shall obtain new ones. In particular, $\{\varphi_3^1,\varphi_3^0\}=\frac{3}{2}\varphi_4^0$, where

$$\varphi_4^0=u_4+2uu_3+\frac{1}{2}(10u_1+3u^2)u_2+3uu_1^2+\frac{1}{2}u^3u_1,$$

while $\{\varphi_2^2,\varphi_4^0\}=4\varphi_4^1$, where

$$\begin{aligned}\varphi_4^1=tu_4+\frac{1}{2}(x+4tu)u_3+\frac{1}{4}(5+6tu^2+3xu+10tu_1)u_2\\ +\frac{1}{8}(12u+3xu^2+4tu^3)u_1+\frac{3}{4}(x+4tu)u_1^2+\frac{1}{8}u^3.\end{aligned}$$

Thus, the action of the symmetries φ_3^1 and φ_2^2 allows one to generate new symmetries of higher order. Symmetries that act in such a way are called *master symmetries* (see [37, 38]). Using these symmetries makes it possible to give a complete description of the Lie algebra $\operatorname{sym}(\mathscr{E})$ in the case of the Burgers equation and of many other equations (especially if we allow for nonlocal master symmetries), [75, 140]

Example 5.6 The Jacobi brackets of the symmetries presented in Example 5.2 for the Gibbons-Tsarev equation are

$$\begin{aligned}\{\varphi_{-4},\varphi_{-3}\}=0,\ \{\varphi_{-4},\varphi_{-2}\}=0,\ &\{\varphi_{-4},\varphi_{-1}\}=0,\qquad \{\varphi_{-4},\varphi_0\}=-2\varphi_{-4},\\ \{\varphi_{-3},\varphi_{-2}\}=0,\ &\{\varphi_{-3},\varphi_{-1}\}=2\varphi_{-4},\ \{\varphi_{-3},\varphi_0\}=-\tfrac{3}{2}\varphi_{-3},\\ &\{\varphi_{-2},\varphi_{-1}\}=\varphi_{-3},\ \ \{\varphi_{-2},\varphi_0\}=-\varphi_{-2},\\ &\{\varphi_{-1},\varphi_0\}=-\tfrac{1}{2}\varphi_{-1}.\end{aligned}$$

Note that if we change the basis by setting

$$\varphi_0\mapsto\varphi_0,\qquad \varphi_{-1}\mapsto-\frac{1}{2}\varphi_{-1}$$

then the brackets will become

$$\{\varphi_i,\varphi_j\}=(j-i)\varphi_{i+j},$$

where, by definition, $\varphi_i = 0$ for $i < -4$. This may be reformulated as follows. Consider the *Witt algebra* $\mathfrak{W}$ generated by the vector fields

$$\mathbf{v}_i = z^{i+1}\frac{\partial}{\partial z}$$

and consider the subspaces $\mathfrak{W}_k^- \subset \mathfrak{W}$ spanned by all $\mathbf{v}_i$ with $i \leq k \leq 0$. Then $\mathfrak{W}_0^-$ is a Lie subalgebra in $\mathfrak{W}$, and all $\mathfrak{W}_k^-$ are its ideals. Then

$$\mathrm{sym}(\mathscr{E}) = \mathfrak{W}_0^-/\mathfrak{W}_{-5}^-$$

in this notation.

Example 5.7 Let us now describe the Lie algebra structure of $\mathrm{sym}(\mathscr{E})$ for the UHE (see Example 5.3; see also [13, 14]). One has

$$\begin{aligned}
&\{\upsilon_0, \varphi_{-1}(T)\} = \varphi_{-1}(T), && \{\upsilon_0, \theta_1(X)\} = -\theta_1(X),\\
&\{\theta_0(X), \theta_0(\bar{X})\} = \theta_0(\bar{X}X' - X\bar{X}'), && \{\theta_0(X), \theta_1(\bar{X})\} = \theta_1(\bar{X}X' - X\bar{X}')\\
&\{\varphi_0(T), \varphi_0(\bar{T})\} = \varphi_0(\bar{T}T' - T\bar{T}'), && \{\varphi_0(T), \varphi_{-1}(\bar{T})\} = \varphi_{-1}(\bar{T}T' - T\bar{T}'),
\end{aligned}$$

all other brackets being trivial. To describe the Lie algebra structure in a different way, introduce the notation:

- let $\mathfrak{V}[\rho]$ denote the Lie algebra of vector fields on the line with a distinguished coordinate ρ;
- let $\mathfrak{L}[\rho] = \mathbb{R}[z] \otimes_{\mathbb{R}} \mathfrak{V}[\rho]$ be the *loop algebra* defined by the relations $[z^i \otimes v, z^j \otimes w] = z^{i+j} \otimes [v, w]$;
- let $\mathfrak{L}_k^+[\rho]$ denote the Lie algebra spanned by the elements $p(z) \otimes v$, $v \in \mathfrak{V}[\rho]$, $p(z) \in \mathbb{R}[z]/(z^k)$ being a truncated polynomial; define also $\mathfrak{L}_{-k}^-[\rho]$, $p(z) \in \mathbb{R}[z^{-1}]/(z^{-k})$, in a similar way;
- finally, similar to Example 5.6, consider subalgebras $\mathfrak{W}_k^+ \subset \mathfrak{W}$, $k \geq 0$, spanned by the elements υ_i, $i \geq k$.

Note that there exist natural actions of $\mathfrak{W}$ on $\mathfrak{L}[\rho]$, of $\mathfrak{W}_l^-$ on $\mathfrak{L}_k^-[\rho]$, and of $\mathfrak{W}_l^+$ on $\mathfrak{L}_k^+[\rho]$. Then we have the following.

Proposition 5.1 *For the universal hierarchy equation $\mathscr{E}$, one has*

$$\mathrm{sym}(\mathscr{E}) \simeq \bar{\mathfrak{W}}_0 \ltimes (\mathfrak{L}_1^+[x] \oplus \mathfrak{L}_{-1}^-[t]),$$

where $\bar{\mathfrak{W}}_0 = \mathfrak{W}_0^+/\mathfrak{W}_1^+ = \mathfrak{W}_0^-/\mathfrak{W}_{-1}^-$.

Example 5.8 The Jacobi brackets of symmetries of the Gibbons-Tsarev equation written in the two-component form (Example 5.4) are

$$\begin{aligned}\{\varphi_{-3},\varphi_{-2}\} = 0,\ \{\varphi_{-3},\varphi_{-1}\} = 0, \quad \{\varphi_{-3},\varphi_0\} &= -3\varphi_{-3},\\ \{\varphi_{-2},\varphi_{-1}\} = \varphi_{-3},\ \{\varphi_{-2},\varphi_0\} &= -2\varphi_{-2},\\ \{\varphi_{-1},\varphi_0\} &= -\varphi_{-1},\end{aligned}$$

and thus $\mathrm{sym}(\mathscr{E}) \simeq \mathfrak{W}_0^-/\mathfrak{W}_{-4}^-$.

5.1.3 Reductions and Invariant Solutions

Consider an equation $\mathscr{E} \subset J^\infty(n,m)$ and its symmetry $\varphi = (\varphi^1,\dots,\varphi^m) \in \mathrm{sym}(\mathscr{E})$. Then the evolution u_τ of a solution $u = (u^1,\dots,u^m)$, $u^j = u^j(x^1,\dots,x^m)$, governed by this symmetry, is described by the equation:

$$\frac{\partial u}{\partial \tau} = \varphi|_{j_\infty(u)}, \qquad u_\tau|_{\tau=0} = u,$$

τ being a parameter of evolution. Fixed points of this evolution are called φ-*invariant solutions*. Thus, invariant solutions satisfy the equation

$$\varphi|_{j_\infty(u)} = 0$$

together with the initial equation $\mathscr{E}$. The obtained system is called φ-*reduction* of $\mathscr{E}$. In a similar way, if $\mathfrak{g} \subset \mathrm{sym}(\mathscr{E})$ is a Lie subalgebra, one can consider $\mathfrak{g}$-reductions.

Two types of reductions are popular in applications:

- assume that $\mathscr{E}$ is an equation in independent variables t and $x^1,\dots,x^n$ and admits t- and x^i-translations. Then $\mathfrak{g}$-invariant solutions, where $\mathfrak{g}$ is an Abelian Lie algebra generated by the translations $\varphi_i = k_i u_t + \omega u_{x^i}$, ω, $k_i \in \mathbb{R}$, are called *traveling waves*;
- let $\mathscr{E}$ admit a scaling symmetry φ. Then φ-invariant solutions are called *self-similar* ones.

Example 5.9 Solutions of the Burgers equation invariant with respect to the symmetry

$$\varphi = ku_t + \omega u_x, \qquad k,\omega \in \mathbb{R},$$

(traveling-wave solutions) are of the form $u = u(\xi)$, where $\xi = \omega t - kx$. Substituting u to the Burgers equation, we obtain the ordinary differential equation

$$k^2 u'' = (\omega + ku)u', \tag{5.4}$$

which is the φ-reduction of the Burgers equation (here "prime" denotes $du/d\xi$). Solving (5.4), we obtain three types of traveling-wave solutions to the Burgers equation:

$$u(\xi) = \alpha \frac{e^{-\frac{\alpha(\xi+\beta)}{k}} + 1}{e^{\frac{-\alpha(\xi+\beta)}{k}} - 1} - \frac{\omega}{k},$$

$$u(\xi) = \alpha \tan\left(\frac{\alpha(\xi+\beta)}{2k}\right) - \frac{\omega}{k}$$

and

$$u(\xi) = -\frac{\xi\omega + 2k^2 + \beta\omega}{k(\xi+\beta)},$$

where α and β are arbitrary constants (we assume that $k \neq 0$).

Example 5.10 Let us now study self-similar solutions of the Burgers equation. The scaling symmetry is $\varphi = u + xu_x + 2tu_t$, and the invariance condition is

$$u + xu_x + 2tu_t = 0.$$

Consequently, self-similar solutions are of the form

$$u = \frac{v(\xi)}{x}, \qquad \xi = \frac{x^2}{t}.$$

Substituting in the Burgers equation, we obtain the φ-reduction in the form

$$4\xi^2 v'' + \xi(\xi + 2v - 2)v' + 2v - v^2 = 0.$$

Generic solutions of this equation form a two-parameter family expressed in terms of *Kummer functions* (see [4]). There also exists a one-parameter family of particular solutions

$$h(\xi) = \frac{2\sqrt{\xi}e^{-\frac{\xi}{4}}}{\sqrt{\pi}\,\mathrm{erf}\left(\frac{\sqrt{\xi}}{2}\right) + \alpha}$$

where

$$\mathrm{erf} = \frac{2}{\sqrt{\pi}} \int_0^x e^{-t^2}\, dt$$

is the *error function* and $\alpha = \mathrm{const}$.

Example 5.11 Consider the KdV equation $u_t = uu_x + u_{xxx}$, and, to make presentation of the final result shorter, transform it to the form

$$u_t = 6uu_x + u_{xxx}$$

by rescaling $u \mapsto 6u$. Let us consider the symmetry $\varphi = cu_x - u_t$. Then the invariant solutions are of the form $u = u(\xi)$, where $\xi = ct + x$ and must satisfy the equation

$$cu' = uu' + u''',$$

which is the φ-reduction of the KdV equation. General 3-parametric solution is expressed by means of *Jacobi elliptic functions*, [4], but physically meaningful solutions constitute a one-parameter family

$$u(\xi) = \frac{c}{2\cosh\left(\frac{\sqrt{c}}{2}(\alpha + \xi)\right)^2}, \qquad \alpha = \text{const.}$$

These are (one-) *soliton solutions* of the KdV equation.

Example 5.12 The scaling symmetry of the KdV equation is

$$\varphi = 3tu_t + xu_x + 2u.$$

Any self-similar solution is of the form

$$u = \frac{f(\xi)}{x^2}, \qquad \xi = \frac{t}{x^3},$$

where the function f satisfies the equation

$$27\xi^3 f''' + 162\xi^2 f'' + (1 + 186\xi + 3\xi f)f' + 2f^2 + 24f = 0,$$

which is the φ-reduction of the KdV equation.

Example 5.13 Two-dimensional symmetry reductions of the universal hierarchy equation $u_{yy} = u_t u_{xy} - u_y u_{tx}$ were completely described in [10]. We shall discuss some interesting cases here. As it follows from Example 5.3, a general invariant solution is determined by the equation

$$\alpha(yu_y + u) + X_1 u_x - X_1' u + T_1 u_t + T_1' y u_y + T_2 u_y + X_2 = 0,$$

where $\alpha = \text{const}$, $X_i = X_i(x)$, and $T_i = T_i(t)$. In particular, in the case $\alpha = 1$, $X_2 = x$, $X_1 = T_1 = T_2 = 0$, we obtain the reduction:

$$vv_{\xi\tau} - v_\xi v_\tau = 2v,$$

which can be transformed to the *Liouville equation* (see, e.g., [48])

$$v_{\xi\tau} = e^{-v}$$

by the change of variables $v \mapsto 2e^v$. When $\alpha = 0$, $X_1 = X_2 = x$, $T_1 = 1$, $T_2 = 0$, the reduction is

$$v_\xi v_{\tau\tau} - v_\tau v_{\xi\tau} = v_{\xi\xi}$$

and is linearizable by the generalized *Legendre transformation* [18]. The case $\alpha = 0$, $X_1 = X_2 = x$, $T_1 = 0$, $T_2 = t$ leads to the equation

$$(1 + \tau v_\tau) v_{\xi\xi} = \tau f_\xi f_{\xi\tau} + f_\xi^2,$$

which can be solved explicitly, etc.

Remark 5.4 It is known ([48]) that the Liouville equation is related to the wave equation $w_{\xi\tau} = 0$ by the family of Bäcklund transformations

$$w_\tau + v_\tau + \lambda e^{\frac{w-v}{2}} = 0, \qquad w_\xi - v_\xi + \frac{2}{\lambda} e^{\frac{w+v}{2}} = 0,$$

where $\lambda \neq 0$ is a parameter.

5.1.4 Nonlocal Symmetries and Shadows

Let $\pi_\infty \colon \mathcal{E} = \{F = 0\} \to M$ be an equation and $\tau \colon \tilde{\mathcal{E}} \to \mathcal{E}$ be a covering given by the conditions (3.4):

$$\frac{\partial w^\alpha}{\partial x^i} = X_i^\alpha,$$

where w^α are nonlocal variables and X_i^α are functions on $\tilde{\mathcal{E}}$. Denote by $\tilde{\pi}_\infty \colon \tilde{\mathcal{E}} \to M$ the composition $\pi_\infty \circ \tau$. A $\tilde{\pi}_\infty$-vertical vector field X on $\tilde{\mathcal{E}}$ is a *nonlocal symmetry* of $\mathcal{E}$ if it preserves the Cartan distribution on $\tilde{\mathcal{E}}$, i.e., if

$$[X, \tilde{\mathcal{C}}] \subset \tilde{\mathcal{C}}. \tag{5.5}$$

In local coordinates, this means that the field X must be of the form

$$\mathbf{E}_{\tilde{\varphi}} = \sum_{\mathbb{I}} \tilde{D}_\sigma(\varphi^j) \frac{\partial}{\partial u_\sigma^j} + \sum_\alpha \psi^\alpha \frac{\partial}{\partial w^\alpha}, \qquad \tilde{\varphi} = (\varphi, \psi^1, \dots, \psi^\alpha, \dots), \tag{5.6}$$

where $\varphi = (\varphi^1, \dots, \varphi^m)$, ψ^α are functions on $\tilde{\mathcal{E}}$, while condition (5.5) amounts to the defining equations

$$\tilde{D}_i(\psi^\alpha) = \tilde{\ell}_{X_i^\alpha}(\varphi) + \sum_\beta \frac{\partial X_i^\alpha}{\partial w^\beta} \psi^\beta, \tag{5.7}$$

$$\tilde{\ell}_{\mathscr{E}}(\varphi) = 0. \tag{5.8}$$

In Eqs. (5.6), (5.7), and (5.8), as before, the subscript $\mathbb{I}$ means summation over internal coordinates, while "tilde" denotes the lift of a $\mathscr{C}$-differential operator to the covering equation.

A vector function $\varphi = (\varphi^1, \dots, \varphi^m)$ that satisfies Eq. (5.8) is called a nonlocal *shadow* in the covering τ. In particular, local symmetries can be regarded as shadows in any covering. Invariantly, a shadow is a derivation

$$\mathbf{E}_\varphi : \mathscr{F}(\mathscr{E}) \to \mathscr{F}(\tilde{\mathscr{E}}), \qquad \mathbf{E}_\varphi(fg) = g\mathbf{E}_\varphi(f) + f\mathbf{E}_\varphi(g), \quad f, g \in \mathscr{F}(\mathscr{E}),$$

that complies with the Cartan connections, i.e., such that the diagram

$$\begin{array}{ccc} \mathscr{F}(\mathscr{E}) & \xrightarrow{\mathbf{E}_\varphi} & \mathscr{F}(\tilde{\mathscr{E}}) \\ {\scriptstyle \mathscr{C}_Z}\big\downarrow & & \big\downarrow{\scriptstyle \tilde{\mathscr{C}}_Z} \\ \mathscr{F}(\mathscr{E}) & \xrightarrow{\mathbf{E}_\varphi} & \mathscr{F}(\tilde{\mathscr{E}}) \end{array}$$

is commutative for any vector field Z on M.

Of course, for any nonlocal symmetry $\mathbf{E}_{\tilde{\varphi}}$, its restriction $\mathbf{E}_\varphi = \mathbf{E}_{\tilde{\varphi}}\big|_{\mathscr{F}(\mathscr{E})}$ is a shadow. A natural question arises: given a shadow $X = \mathbf{E}_{\tilde{\varphi}}$, does there exist a nonlocal symmetry $\tilde{X}$ with this shadow? In other words, for a given solution of Eq. (5.8), do there exist functions ψ^α satisfying Eq. (5.7)? If such a symmetry exists, it is called a *lift* of X. The answer is as follows: there exists a covering $\tilde{\tau}: \tilde{\tilde{\mathscr{E}}} \to \tilde{\mathscr{E}}$ (uniquely defined up to an equivalence) and a shadow $\tilde{X}$ in this covering such that $\tilde{X}\big|_{\mathscr{F}(\mathscr{E})} = X$ (see [66]).

The shadow $\tilde{X}$ is not unique, but is defined up to *invisible symmetries*: a nonlocal symmetry $\mathbf{E}_{\tilde{\varphi}}$, $\tilde{\varphi} = (\varphi, \psi^1, \dots, \psi^\alpha, \dots)$ is called invisible if it is τ-vertical, i.e., if its φ-component vanishes. The defining equations for invisible symmetries are

$$\tilde{D}_i(\psi^\alpha) = \sum_\beta \frac{\partial X_i^\alpha}{\partial w^\beta} \psi^\beta \tag{5.9}$$

for all $\alpha, \beta = 1, \dots$

Consider a particular case of the above constructions when $\tau = \tau_\omega$ is the Abelian covering associated with a nontrivial two-component conservation law:

$$\omega = (X_1\, dx^1 + X_2\, dx^2) \wedge dx^3 \wedge \dots \wedge dx^n$$

(see Sect. 3.1.2). This covering is endowed with the nonlocal variables w^ρ, where ρ is a multi-index consisting of the integers $3, \dots, n$ and is described by the system

$$w^\rho_{x^1} = D_\rho(X_1), \qquad w^\rho_{x^2} = D_\rho(X_2), \qquad w^\rho_{x^i} = w^{\rho i}, \quad i > 2.$$

Nonlocal symmetries are

$$\mathbf{E}_{\tilde{\varphi}} = \sum_{\mathbb{I}} \tilde{D}_\sigma(\varphi^j)\frac{\partial}{\partial u^j_\sigma} + \sum_\rho \tilde{D}_\rho(\psi)\frac{\partial}{\partial w^\rho}, \qquad \tilde{\varphi} = (\varphi^1, \dots, \varphi^m, \psi),$$

in this case, where the vector function $\tilde{\varphi}$ enjoys the system

$$\tilde{D}_i(\psi) = \tilde{\ell}_{X_i}(\varphi), \quad i = 1, 2, \qquad \tilde{\ell}_{\mathcal{E}}(\varphi) = 0.$$

In particular, invisible symmetries correspond to vector functions $\tilde{\varphi} = (0, \dots, 0, \psi)$ satisfying the equations

$$\tilde{D}_1(\psi) = \tilde{D}_2(\psi) = 0,$$

i.e., are of the form

$$\mathbf{E}_\psi = \sum_\rho \tilde{D}_\rho(\psi)\frac{\partial}{\partial w^\rho},$$

where $\psi = \psi(x^3, \dots, x^n)$ is an arbitrary smooth function. In particular, $\psi = \text{const}$ when $n = 2$

Example 5.14 Consider the Abelian covering of the Burgers equation associated with the conservation law $\omega = u\,dx + \left(\frac{u^2}{2} + u_x\right)dt$. Nonlocal symmetries in this covering are vector functions $\tilde{\varphi} = (\varphi, \psi)$ depending on x, t, u_k, $k = 0, 1, \dots$, and w and satisfying the system

$$\tilde{D}_x(\psi) = \varphi \qquad \tilde{D}_t(\psi) = u\varphi + \tilde{D}_x(\varphi), \quad \tilde{D}_t(\varphi) = u_1\varphi + u\tilde{D}_x(\varphi) + \tilde{D}_x^2(\varphi), \tag{5.10}$$

where

$$\tilde{D}_x = D_x + u\frac{\partial}{\partial w}, \qquad \tilde{D}_t = D_t + \left(\frac{2}{2}u^2 + u_1\right)\frac{\partial}{\partial w}.$$

Due to the first equation in (5.10), any symmetry is completely determined by its ψ-component

Direct computations for order ≤ 1 lead to the following results:

$$\begin{aligned} \psi_f &= f(t, x)e^{-\frac{1}{2}w}, \\ \psi_{\text{inv}} &= 1, \\ \psi_1^0 &= u, \end{aligned}$$

$$\psi_1^1 = tu + x,$$

$$\psi_2^0 = \frac{1}{2}u^2 + u_x,$$

$$\psi_2^1 = tu_x + \frac{1}{2}xu + \frac{1}{2}u^2t,$$

$$\psi_2^2 = t^2u_x + tux + \frac{1}{2}t^2u^2 + \frac{1}{2}x^2 + t,$$

where $f(t,x)$ is an arbitrary solution of the heat equation $f_t = f_{xx}$.

Let us compare these computations with the results of Example 5.1. Obviously, the symmetries ψ_i^j are the lifts of φ_i^j to the covering under consideration. The symmetry ψ_{inv} is invisible, while φ_f is the lift of the nonlocal shadow

$$\varphi_f = \left(f_x - \frac{1}{2}fu\right)e^{-\frac{1}{2}w},$$

where f is a solution of the heat equation.

Example 5.15 Computations of nonlocal symmetries of order ≤ 5 in the covering over the KdV equation associated with the conservation law

$$\omega = u\,dx + \left(\frac{u^2}{2} + u_{xx}\right)dt$$

reveal, in comparison with the local case, only one new symmetry $\partial/\partial w$, which is invisible. But computation of shadows gives a nontrivial result:

$$\varphi_5^1 = (tu_5 + uu_3 + u_1u_2 + u^2u_1) + x(u_3 + uu_1) + \frac{4}{3}u_2 + \frac{4}{9}u^2 + \frac{1}{9}u_1w.$$

This shadow cannot be lifted.

Example 5.16 To describe nonlocal symmetries in the covering

$$u_x = v + w, \qquad u_y = -vw$$

of the scalar Gibbons-Tsarev equation

$$u_{yy} = u_yu_{xx} - u_xu_{xy} + 1$$

over its two-component version

$$v_y + wv_x = \frac{1}{w - v}, \qquad w_y + vw_x = \frac{1}{v - w},$$

it is convenient to compare Examples 5.2 and 5.4. It can be seen that the symmetries $\varphi_{-3}, \dots, \varphi_0$ in Example 5.4 are the lifts of the symmetries from Example 5.2 with the same notation, while the symmetry φ_{-1} is invisible in this covering.

Example 5.17 We shall now study nonlocal symmetries in the positive and negative coverings associated with the Lax pair of the UHE (see Example 3.10).

Let us begin with the positive covering. The following facts are valid:

- All the local symmetries υ_0, $\theta_0(X)$, $\theta_1(X)$, $\varphi_{-1}(T)$, and $\varphi_0(T)$ (see Example 5.3) can be lifted to the covering.
- Direct computations reveal two nonlocal shadows

$$\upsilon_1 = 2q_1 u_y - y u_t, \qquad \upsilon_2 = -3q_2 u_y + 2q_1 u_t - y u_y q_{1,t}$$

  of weights 1 and 2, respectively, and both can be lifted to nonlocal symmetries.
- There exists a "negative" hierarchy of invisible symmetries

$$\varphi_{-k}(T) = \varphi^0(T)\frac{\partial}{\partial q_{k-1}} + \varphi^1(T)\frac{\partial}{\partial q_k} + \dots + \varphi^i(T)\frac{\partial}{\partial q_{i+k-1}} + \dots, \quad k \geq 2,$$

  where

$$\varphi^0(T) = T, \quad \varphi^i(T) = \frac{1}{i!}\mathscr{Y}^i(\varphi^0), \qquad \mathscr{Y} = -y\frac{\partial}{\partial t} + 2q_1\frac{\partial}{\partial y} + \sum_{\alpha\geq 1}(i+1)q_{i+1}\frac{\partial}{\partial q_i}.$$

  One has the weights $[\varphi_{-k}(T)] = -k$.
- The action of υ_1 generates two additional "positive" hierarchies defined by

$$\upsilon_k = \{\upsilon_1, \upsilon_{k-1}\}, \quad \varphi_l(T) = \{\upsilon_1, \varphi_{l-1}(T)\}, \qquad k \geq 3, \quad l \geq 1.$$

Using notation of Example 5.7, the Lie algebra structure of nonlocal symmetries is described by

Proposition 5.2 *For the Lie algebra* $\mathrm{sym}_{\tau^+}(\mathscr{E})$ *of the universal hierarchy equation in the positive covering, one has the following isomorphism:*

$$\mathrm{sym}_{\tau^+}(\mathscr{E}) \simeq \mathfrak{W}_0^+ \ltimes (\mathfrak{L}_2^+[x] \oplus \mathfrak{L}[t]),$$

where semi-direct product is defined by the natural actions described in Example 5.7.

In the case of the negative covering, the situation is as follows:

- Similar to the positive case, all local symmetries can be lifted to the covering.
- By direct computations, one can construct two nonlocal shadows

$$\upsilon_{-1} = 2r_1 - uu_x, \qquad \upsilon_{-2} = 3r_2 - 2r_1 u_x - ur_{1,x} + uu_x^2$$

  with $[\upsilon_i] = i$ and lift them to the covering.

- There exists a "positive" hierarchy of invisible symmetries

$$\theta_k(X) = \theta^0(X)\frac{\partial}{\partial r_{k-1}} + \theta^1(X)\frac{\partial}{\partial r_k} + \cdots + \theta^i(X)\frac{\partial}{\partial r_{i+k-1}} + \ldots, \qquad k \geq 2,$$

where

$$\theta^0(X) = X, \quad \varphi^i(T) = \frac{1}{i!}\mathscr{Y}^i(\theta^0), \qquad \mathscr{Y} = u\frac{\partial}{\partial x} + 2r_1\frac{\partial}{\partial u} + \sum_{\alpha \geq 1}(i+1)r_{i+1}\frac{\partial}{\partial r_i}.$$

- The action of υ_{-1} generates two additional "negative" hierarchies defined by

$$\upsilon_k = \{\upsilon_{-1}, \upsilon_{k-1}\}, \quad \theta_l(T) = \{\upsilon_{-1}, \theta_{l-1}(T)\}, \qquad k \leq -3, \quad l \leq 1.$$

The Lie algebra structure in thus obtained space of nonlocal symmetries is given by the following.

Proposition 5.3 *For the Lie algebra* $\mathrm{sym}_{\tau^-}(\mathscr{E})$ *of the universal hierarchy equation in the negative covering, one has the following isomorphism:*

$$\mathrm{sym}_{\tau^-}(\mathscr{E}) \simeq \mathfrak{W}_0^- \ltimes (\mathfrak{L}_2^-[t] \oplus \mathfrak{L}[x]),$$

where semi-direct product is defined by the natural actions described in Example 5.7.

5.2 Examples

We now pass to the CDE implementation of symmetry computations.

5.2.1 *Korteweg-de Vries Equation*

The determining equations of symmetries can be solved by using an ansatz from the weight approach or by CRACK. Let us start with the weight approach; we describe the program file kdv_hsy1.red.

The defining condition for symmetries of the KdV equation is

$$D_t(\varphi) = u_x\varphi + uD_x(\varphi) + D_x^3(\varphi).$$

The functions $\varphi_1 = u_x$ and $\varphi_2 = uu_x + u_{xxx}$ are symmetries of the KdV equation (x- and t-translations, respectively). To compute them, load in CDE the KdV equation as in Example 3.2.1, and make an ansatz for the unknown symmetry sym as a weighted polynomial whose monomials have weights ranging from 0 to 8:

```
ctel:=0;
operator c,equ;
l_grad_mon:=der_deg_ordering(0,all_parametric_der);
gradmon:=graded_mon(1,8,l_grad_mon);
gradmon:={1} .gradmon;
ansatz:=for each el in gradmon join el;
sym:=(for each el in ansatz sum (c(ctel:=ctel+1)*el));
```

Then, we solve the equation $\ell_{\mathrm{KdV}}(\varphi) = 0$:

```
mk_cdiffop(ell_kdv,1,1);
for all phi let ell_kdv(1,1,phi)
= td(phi,t)-u*td(phi,x)-u_x*phi-td(phi,x,3);

equ 1:=ell_kdv(1,1,sym);
```

The above equation is a polynomial with respect to all variables; we obtain a sparse system of linear algebraic equations by the requirement that all coefficients of the polynomial should vanish. The unknowns are the constants c(1),c(2),.... The solver is programmed as in Example 3.2.1. The result is

```
(5*c(12)*u**2*u_x + 10*c(12)*u*u_3x +
20*c(12)*u_2x*u_x + 6*c(12)*u_5x +
5*c(6)*u*u_x + 5*c(6)*u_3x + 5*c(3)*u_x)/5;
```

Note the first higher symmetry, corresponding to the constant c(12).

We can use CRACK as well for the same purpose. This time the program file is kdv_hsy2.red; the equation to be solved is just the same. The difference is that we assume that sym be a function of dependent variables of order (of derivatives) up to 5:

```
ansatz_vars:=for i:=0:5 join
  selectvars(0,i,dep_var,all_parametric_der);
for each el in ansatz_vars do depend sym,el;
```

Then we introduce the variables with respect to which the system is polynomial, and collect their coefficients as new equations:

```
split_vars:=
  cde_difflist(all_parametric_der,ansatz_vars);
system_eq:=splitvars_list(lin_eq,split_vars);
```

Finally we start CRACK on the system of equations system_eq:

```
load_package crack;
unk:={sym};
crack_sol:=crack(system_eq,{},unk,{});
```

The solution is

```
crack_sol:={{{},
{sym=(5*c_14*u**2*u_x + 10*c_14*u*u_3x +
20*c_14*u_2x*u_x + 6*c_14*u_5x +
6*c_30*u_x + 6*c_32*u*u_x + 6*c_32*u_3x)/6},
{c_14,c_32,c_30},
{}}};
```

where c_14,c_30,c_32 are constants.

The conservation law $\omega_1 = u\,dx + \left(\frac{u^2}{2} + u_2\right) dt$ from Sect. 3.2.1 defines the covering

$$w_x = u, \qquad w_t = \frac{u^2}{2} + u_{xx}$$

and

$$\mathbf{E}_{\varphi_1} + u\frac{\partial}{\partial w}, \qquad \mathbf{E}_{\varphi_2} + \left(\frac{u^2}{2} + u_2\right)\frac{\partial}{\partial w}$$

are nonlocal symmetries in this covering. Computing them is the same as computing symmetries for any equation. It is perhaps more interesting to compute shadows of symmetries (program file kdv_hsy3.red), introducing the new equation

```
principal_der:={u_t,w_t,w_x};
de:={u*u_x+u_3x,u**2/2 + u_2x,u};
```

We form an ansatz for CRACK:

```
ansatz_vars:=for i:=0:5 join
  selectvars(0,i,dep_var,all_parametric_der);
for each el in ansatz_vars do depend sym,el;
for each el in indep_var do depend sym,el;
split_vars:=
  cde_difflist(all_parametric_der,ansatz_vars);
unk:={sym};
lin_eq:={ell_kdv(1,1,sym)};
system_eq:=splitvars_list(lin_eq,split_vars);
load_package crack;
crack_sol:=crack(system_eq,{},unk,{});
```

The result is

```
crack_sol:={{{},
{sym=( - 135*c_64*t*u**2*u_x - 270*c_64*t*u*u_3x
 - 540*c_64*t*u_2x*u_x - 162*c_64*t*u_5x
 - 72*c_64*u**2 - 54*c_64*u*u_x*x - 216*c_64*u_2x
 - 54*c_64*u_3x*x - 18*c_64*u_x*w - 2*c_68*u_x
```

```
 - 2*c_69*t*u_x - 2*c_69 - 12*c_70*u*u_x
 - 12*c_70*u_3x - 12*c_72*t*u*u_x - 12*c_72*t*u_3x
 - 8*c_72*u - 4*c_72*u_x*x - 30*c_73*u**2*u_x
 - 60*c_73*u*u_3x - 120*c_73*u_2x*u_x
 - 36*c_73*u_5x)/36},
{c_64,c_69,c_68,c_72,c_70,c_73},
{}}};
```

The shadow corresponding to the constant c_64 is nontrivial, as it contains *w*.

5.2.2 Burgers Equation

In this Section we will compute some higher symmetries of the Burgers equation and their Jacobi bracket. We begin by the computation of higher symmetries (see the example file bur_hsy1.red). The weights are

```
deg_indep_var:={-1,-2};
deg_dep_var:={1};
```

We prepare the ansatz as follows:

```
ctel:=0;
operator c,equ;
l_grad_var:=der_deg_ordering(0,all_parametric_der);
gradmon:=graded_mon(1,10,l_grad_var);
gradmon:={1} . gradmon;
ansatz:=for i:=0:3 join
  mkallgradmon_evenind(1,gradmon,i);
sym:=for each el in ansatz sum
  (c(ctel:=ctel+1)*el);
```

and we solve the equation

```
equ 1:=td(sym,t) - td(sym,x,2) - 2*u*td(sym,x)
  - 2*u_x*sym;
```

We obtain

```
sym := (6*c(34)*u*u_x + 3*c(34)*u_2x
  + 16*c(22)*t*u**3*u_x + 24*c(22)*t*u**2*u_2x
  + 16*c(22)*t*u*u_3x + 48*c(22)*t*u*u_x**2
  + 40*c(22)*t*u_2x*u_x + 4*c(22)*t*u_4x
  + 2*c(22)*u**3 + 6*c(22)*u**2*u_x*x
  + 6*c(22)*u*u_2x*x - c(22)*u_2x + 2*c(22)*u_3x*x
  + 6*c(22)*u_x**2*x + 6*c(20)*u_x
  + 18*c(12)*t*u**2*u_x + 18*c(12)*t*u*u_2x
  + 6*c(12)*t*u_3x + 18*c(12)*t*u_x**2
```

```
+ 3*c(12)*u**2 + 6*c(12)*u*u_x*x
+ 3*c(12)*u_2x*x + 24*c(6)*t*u*u_x
+ 12*c(6)*t*u_2x + 6*c(6)*u + 6*c(6)*u_x*x
+ 6*c(3)*t*u_x + 3*c(3))/6;
```

Similar results could be obtained by the CRACK approach; see the file bur_hsy2.red.

Then, we continue the computation by loading the above results in the file bur_jbr1.red. We declare four evolutionary vector fields out of the above symmetries (the names are taken by Example 5.1, evsym01 corresponds to φ_1^0 and so on):

```
mk_evfield(evsym01);
evsym01(0,1):=df(sym,c(20));
evsym01(1,1):=0;

mk_evfield(evsym11);
evsym11(0,1):=df(sym,c(3));
evsym11(1,1):=0;

mk_evfield(evsym02);
evsym02(0,1):=df(sym,c(34));
evsym02(1,1):=0;

mk_evfield(evsym12);
evsym12(0,1):=df(sym,c(6));
evsym12(1,1):=0;

mk_evfield(evsym13);
evsym13(0,1):=df(sym,c(12));
evsym13(1,1):=0;
```

The command mk_evfield(evfield) declares a supervectorfield φ of evolutionary type whose representation in Reduce is evfield. This means that the vector field is of type

$$\varphi = \varphi_0^i \frac{\partial}{\partial u^i} + \varphi_1^j \frac{\partial}{\partial p^j},$$

where the components φ_0^i are accessed by the command evfield(0,i) and the components φ_1^j are accessed by the command evfield(1,j).

A nontrivial bracket is

```
jacobi_bracket(evsym01,evsym12,ev_01_12);

ev_01_12(0,1);

- u_x;
```

The command `jacobi_bracket(evsym01,evsym12,ev_01_12)` takes two evolutionary supervectorfields, computes the Jacobi bracket, and stores the result in the third supervectorfield `ev_01_12`. More nontrivial brackets are generated by `evsym13`.

5.2.3 Dispersionless Boussinesq System

The functions

$$\varphi_1 = (0, 1, 0), \quad \varphi_2 = (u_1, v_1, w_1), \quad \varphi_3 = (w_1 w + v_1, -w_1 u - 3u_1 w, u_1)$$

are local symmetries of this system. The conservation laws from Sect. 3.2.2 allow to introduce the nonlocal variables p_1, p_2, p_3 defined by

$$\begin{aligned} (p_1)_x &= w, & (p_1)_t &= u, \\ (p_2)_x &= u, & (p_2)_t &= v + \frac{w^2}{2}, \\ (p_3)_x &= v + w^2, & (p_3)_t &= -uw. \end{aligned}$$

Then $\varphi = (\varphi^1, \varphi^2, \varphi^3)$, where

$$\begin{aligned} \varphi^1 &= -20p_3 u_1 - 12p_2(w_1 w + v_1) + 6p_1(w_1 u + u_1 w) + 8u(3v + w^2), \\ \varphi^2 &= -20p_3 v_1 + 12p_2(w_1 u + 3u_1 w) + 6p_1(3w_1 w^2 + u_1 u + 2v_1 w) \\ &\quad - 39u^2 w + 16v^2 - 12w^4, \\ \varphi^3 &= -20p_3 w_1 - 12p_2 u_1 - 6p_1(2w_1 w + v_1) + 9u^2 + 16vw + 16w^3, \end{aligned}$$

is a nonlocal shadow in this setting.

The above shadow can be easily obtained using a weight approach (see the program file bou_sh1.red). Let us load the covering:

```
indep_var:={x,t};
dep_var:={u,v,w,z1,z2,z3};
odd_var:={p,q,r};
deg_indep_var:={-1,-2};
deg_dep_var:={3,4,2,1,2,3};
deg_odd_var:={0,0,0};
total_order:=8;
```

Note the weight of the nonlocal variables. The equation is

```
principal_der:={u_t,v_t,w_t,
  z1_x,z1_t,
```

```
    z2_x,z2_t,
    z3_x,z3_t
      };
de:={w*w_x + v_x, - u*w_x - 3*w*u_x,u_x,
    w,u,
    u,v+(w**2/2),
    v+w**2,-u*w
    };
```

After loading the linearization

```
mk_superfun(lbou_sf,1,3);
mk_superfun(lbou_star_sf,1,3);
in "bou_ell1_res.red";
conv_superfun2cdiff(lbou_sf,lbou);
```

we initialize the space of weighted monomials as in the previous examples and assume a generating function with components of weights ≤ 8:

```
phiu:=for each el in ansatz sum c(ctel:=ctel+1)*el;
phiv:=for each el in ansatz sum c(ctel:=ctel+1)*el;
phiw:=for each el in ansatz sum c(ctel:=ctel+1)*el;
unk:={phiu,phiv,phiw};
```

Then we define the equations

```
nc:=3;
equ 1:=for j:=1:nc sum lbou(1,j,part(unk,j));
equ 2:=for j:=1:nc sum lbou(2,j,part(unk,j));
equ 3:=for j:=1:nc sum lbou(3,j,part(unk,j));
```

and solve them for the coefficients c as usual. Note that the more general CRACK approach is less effective in this case: the computation would be much slower. The result of the computation is

```
phiu;

( - 12*c(110)*u*v - 4*c(110)*u*w**2
- 3*c(110)*u*w_x*z1 - 3*c(110)*u_x*w*z1
+ 10*c(110)*u_x*z3 + 6*c(110)*v_x*z2
+ 6*c(110)*w*w_x*z2 + 6*c(70)*u*w_x
+ 6*c(70)*u_x*w + 6*c(36)*v_x
+ 6*c(36)*w*w_x + 6*c(22)*u_x)/6;

phiv;

(12*c(269) + 39*c(110)*u**2*w - 6*c(110)*u*u_x*z1
- 12*c(110)*u*w_x*z2 - 36*c(110)*u_x*w*z2
- 16*c(110)*v**2 - 12*c(110)*v_x*w*z1
```

```
+ 20*c(110)*v_x*z3 + 12*c(110)*w**4
- 18*c(110)*w**2*w_x*z1 + 12*c(70)*u*u_x
+ 24*c(70)*v_x*w + 36*c(70)*w**2*w_x
- 12*c(36)*u*w_x - 36*c(36)*u_x*w
+ 12*c(22)*v_x)/12;

phiw;

( - 9*c(110)*u**2 + 12*c(110)*u_x*z2
- 16*c(110)*v*w + 6*c(110)*v_x*z1
- 16*c(110)*w**3 + 12*c(110)*w*w_x*z1
+ 20*c(110)*w_x*z3 - 12*c(70)*v_x
- 24*c(70)*w*w_x + 12*c(36)*u_x
+ 12*c(22)*w_x)/12;
```

5.2.4 Camassa-Holm Equation

The functions

$$
\begin{aligned}
\varphi_1 &= u^1_0,\\
\varphi_2 &= u^0_1,\\
\varphi_3 &= tu^0_1 + u^0_0,\\
\varphi_4 &= \alpha u^0_0(3u^0_0 u^1_0 - 4u^0_1) - 2u^2_0(u^0_0 u^1_0 - u^0_1) - (u^1_0)^3 + 2u^1_2,
\end{aligned}
$$

are local symmetries of the Camassa-Holm equation, while

$$\varphi_7 = -2s_1 u^1_0 + t\varphi_4 + 4\big(-\alpha(u^0_0)^2 + u^0_0 u^2_0 + u^1_1\big)$$

is a nonlocal shadow, where the nonlocal variable s_1 is defined by the equations

$$(s_1)_t = u^0_0 u^2_0 + \frac{u^2_1 - 3\alpha(u^0_0)^2}{2}, \qquad (s_1)_x = \alpha u^0_0 - u^2_0,$$

and arises from the conservation laws indicated in Sect. 3.2.3.

The example file ch_hsy1.red contains a program for computing local higher symmetries of the Camassa-Holm equation. We will describe the example file ch_hsy2.red, which is more general than the previous one as it also includes nonlocal variables and allows us to compute shadows of symmetries, besides local higher symmetries. We define the jet space with the nonlocal variables as follows:

```
indep_var:={t,x};
dep_var:={u,s1,s2};
odd_var:={q};
total_order:=8;
```

with consistent weights

```
deg_indep_var:={-2,-1};
deg_dep_var:={1,2,3};
deg_odd_var:={1};
```

The nonlocal variables can be entered as

```
s1x:=alpha*u - u_2x;
s1t:=u*u_2x + (1/2)*(u_x**2 - 3*alpha*(u**2));
s2x:=alpha*(u**2) - u*u_2x;
s2t:= - 2*alpha*(u**3) + 2*(u**2)*u_2x + u*u_tx
   - u_t*u_x;
```

and the equations are loaded by

```
principal_der:={u_3x,s1_x,s1_t,s2_x,s2_t};
de:={(alpha*(u_t + 3*u*u_x) - u_t2x - 2*u_x*u_2x)/u,
   s1x,s1t,s2x,s2t};
```

The linearization of the equation is loaded as in Sect. 2.3.3. The constant `alpha` must be added to the weighted variables in the right place by the command

```
l_grad_mon:=der_deg_ordering(0,all_parametric_der);
l_grad_mon:=part(l_grad_mon,2):=
   alpha . part(l_grad_mon,2);
gradmon:=graded_mon(1,15,l_grad_mon);
gradmon:={1} . gradmon;
```

The most interesting solutions start to appear at weight 6, where we generate the ansatz as

```
grmont:=mkallgradmon_evenind(1,gradmon,6);
sym:=for each el in grmont sum
   (c(ctel:=ctel+1)*el);
```

and the equation as

```
equ 1:=num lch(1,1,sym);
```

Solving as usual (with the switch `coefficient_check` set to `off` in order to consider the most generic case with respect to the value of `alpha`), we obtain

```
sym:=
 (3*c(219)*alpha**2*u_x + 3*c(206)*alpha*u**2*u_x
 - 4*c(206)*alpha*u*u_t - 2*c(206)*u*u_2x*u_x
 + 2*c(206)*u*u_t2x + 2*c(206)*u_2tx
```

```
+ 2*c(206)*u_2x*u_t - c(206)*u_x**3
+ 3*c(21)*alpha**2*t*u**2*u_x
- 4*c(21)*alpha**2*t*u*u_t
- 4*c(21)*alpha**2*u**2 - 2*c(21)*alpha*s1*u_x
- 2*c(21)*alpha*t*u*u_2x*u_x
+ 2*c(21)*alpha*t*u*u_t2x
+ 2*c(21)*alpha*t*u_2tx + 2*c(21)*alpha*t*u_2x*u_t
- c(21)*alpha*t*u_x**3 + 4*c(21)*alpha*u*u_2x
+ 4*c(21)*alpha*u_tx)/3;
```

We stress the fact that obtaining the right combination of weight and algebraic degree of the independent variables is a matter of trial and error, and it can be quite difficult. Given the high number of variables involved, the CRACK approach is not always effective.

5.2.5 *Multi-dimensional Examples*

Let now consider some examples in dimension > 2.

5.2.5.1 Universal Hierarchy Equation

We describe the program uh_sym1.red for computing local symmetries of the universal hierarchy equation (Example 3.10).

After introducing the linearization,

```
mk_cdiffop(ell_uh,1,{1},1);
for all psi let
  ell_uh(1,1,psi) = td(psi,y,2) - u_t*td(psi,x,y)
    + u_y*td(psi,x,t) - u_xy*td(psi,t)
    + u_tx*td(psi,y);
```

we define an ansatz for a symmetry φ. We require φ to be a function of independent variables, u, and its first derivatives.

```
ansatz_vars:=for i:=0:1 join
  selectvars(0,i,dep_var,all_parametric_der);
for each el in ansatz_vars do depend phi,el;
depend phi,t,x,y;
split_vars:=
  cde_difflist(all_parametric_der,ansatz_vars);
unk:={phi};
```

The equation of symmetries is $\ell_{\mathscr{E}}(\varphi) = 0$, where $\mathscr{E}\colon F = 0$ is the universal hierarchy equation. In the program we have

```
lin_eq:={ell_uh(1,1,phi)};
```

The system of equations is generated by all coefficients of the variables with respect to which the expression `lin_eq` is polynomial:

```
system_eq:=splitvars_list(lin_eq,split_vars);
load_package crack;
crack_sol:=crack(system_eq,{},unk,{});
```

We have

```
crack_sol;

{{{},
{psi= - df(c_16,x)*u + df(c_18,t)*u_y*y
  + c_16*u_x + c_18*u_t + c_20 + c_22*u_y
  - c_24*u - c_24*u_y*y},
{c_16,c_18,c_20,c_22,c_24},
{}}};
```

where `c_16,c_20` depend on x only, `c_18,c_22` depend on t only, and `c_24` is a constant. This recovers the symmetries of Example 5.3.

5.2.5.2 Pavlov Equation

Let us consider the Pavlov equation (see [29, 112]):

$$u_{yy} = u_{tx} + u_y u_{xx} - u_x u_{xy}. \tag{5.11}$$

It is an integrable equation as it admits a linear Lax pair. The above equation does not have known higher symmetries; however its cotangent covering (see Chap. 9)

$$\begin{aligned} u_{yy} &= u_{tx} + u_y u_{xx} - u_x u_{xy} \\ v_{yy} &= v_{tx} + u_y v_{xx} - u_x v_{xy} + 2(u_{xy} v_x - u_{xx} v_y) \end{aligned} \tag{5.12}$$

has higher symmetries (see [9] and references therein). This provides a first example of a system of PDEs in dimension higher than $1 + 1$ that admits local higher symmetries. Here we show that the system admits symmetries depending on third-order variables.

Remark 5.5 The fact that Eq. (5.12) admits symmetries depending on higher-order jet variables was observed in [9]. Indeed, the cotangent covering ($\mathcal{T}^*\mathcal{E}$) is always an Euler-Lagrange equation, and thus the spaces $\operatorname{sym}(\mathcal{T}^*\mathcal{E})$ and $\operatorname{cosym}(\mathcal{T}^*\mathcal{E})$ coincide. On the other hand, any cosymmetry of $\mathcal{E}$ can be canonically lifted to a cosymmetry of $\mathcal{T}^*\mathcal{E}$. Consequently, higher-order cosymmetries of $\mathcal{E}$ give rise to higher-order symmetries of $\mathcal{T}^*\mathcal{E}$. For example, the cosymmetry $\psi_{-5} = u_{xxx}$ of the Gibbons-Tsarev equation (see Example 4.2) will generate a third-order symmetry of its cotangent covering.

It is suitable to take functions

$$u^{(0)}_{k,l} = u_{\underbrace{x\ldots x}_{k\text{ times}}\underbrace{t\ldots t}_{l\text{ times}}}, \qquad u^{(1)}_{k,l} = u_{y\underbrace{x\ldots x}_{k\text{ times}}\underbrace{t\ldots t}_{l\text{ times}}}$$

for internal coordinates on $\mathscr{E}$. The total derivatives restricted to $\mathscr{E}$ are

$$D_x = \frac{\partial}{\partial x} + \sum_{k,l}\left(u^{(0)}_{k+1,l}\frac{\partial}{\partial u^{(0)}_{k,l}} + u^{(1)}_{k+1,l}\frac{\partial}{\partial u^{(1)}_{k,l}}\right),$$

$$D_t = \frac{\partial}{\partial t} + \sum_{k,l}\left(u^{(0)}_{k,l+1}\frac{\partial}{\partial u^{(0)}_{k,l}} + u^{(1)}_{k,l+1}\frac{\partial}{\partial u^{(1)}_{k,l}}\right),$$

$$D_y = \frac{\partial}{\partial y} + \sum_{k,l} u^{(1)}_{k,l}\frac{\partial}{\partial u^{(0)}_{k,l}} + \sum_{k,l} D_x^k D_t^l\left(u^{(0)}_{1,1} + u^{(1)}_{0,0}u^{(0)}_{2,0} - u^{0}_{1,0}u^{(1)}_{1,0}\right)\frac{\partial}{\partial u^{(1)}_{k,l}}.$$

Let us load the system (5.12) in cde (program file pav_hsy1.red):

```
indep_var:={t,x,y};
dep_var:={u,v};
odd_var:={p,q};
total_order:=10;
principal_der:={u_2y,v_2y};
de:={u_tx + u_y*u_2x - u_x*u_xy,
  v_tx + u_y*v_2x - u_x*v_xy
  + 2*(u_xy*v_x - u_2x*v_y)};
```

with the following consistent weights

```
deg_indep_var:={-5,-1,-3};
deg_dep_var:={1,1};
deg_odd_var:={0,0};
```

The linearization of the system has been previously computed using the program file pav_ell1.red:

```
mk_superfun(lpav_sf,1,2);
lpav_sf(1):= - p_2x*u_y + p_2y - p_tx + p_x*u_xy
 + p_xy*u_x - p_y*u_2x;
lpav_sf(2):=2*p_2x*v_y + p_x*v_xy - 2*p_xy*v_x
 - p_y*v_2x - q_2x*u_y + q_2y - q_tx
 - 2*q_x*u_xy + q_xy*u_x + 2*q_y*u_2x;
conv_superfun2cdiff(lpav_sf,lpav);
```

Now we shall define an ansatz for a two-component vector function `phiu, phiv` that will be in the kernel of ℓ_F:

```
phiu:=0;
```

We generate a weighted polynomial phiv with weights up to 8:

```
l_grad_mon:=der_deg_ordering(0,all_parametric_der);
gradmon:=graded_mon(1,8,l_grad_mon);
gradmon:={1} . gradmon;
ansatz:=for each el in gradmon join el;
phiv:=for each el in ansatz sum c(ctel:=ctel+1)*el;
unk:={phiu,phiv};
```

Finally, the equations on the coefficients of the polynomial are defined by the coefficients of the following expressions:

```
nc:=length(dep_var);
equ 1:=for j:=1:nc sum lpav(1,j,part(unk,j));
equ 2:=for j:=1:nc sum lpav(2,j,part(unk,j));
```

Solving the system (it has 6961 equations), we obtain

```
phiv := (6*c(487)*u_2x**2*u_x + 6*c(487)*u_2x*u_xy
  + 6*c(487)*u_2xy*u_x + 6*c(487)*u_3x*u_x**2
  + 6*c(487)*u_3x*u_y + 6*c(487)*u_t2x
  + 6*c(143)*u_2x**2 + 12*c(143)*u_2xy
  + 12*c(143)*u_3x*u_x + 3*c(120)*u_t
  + 6*c(120)*u_x**3 + 9*c(120)*u_x*u_y
  + 6*c(38)*u_3x + 6*c(32)*u_x**2 + 4*c(32)*u_y
  + 6*c(8)*u_x + 6*c(2)*v + 6*c(1))/6;
```

which contains some of the higher symmetries found in [9]. In a similar way one could recover other results from [9] concerning more multi-dimensional PDEs.

Chapter 6
The Tangent Covering

Abstract The tangent covering is an equation naturally related to the initial equation $\mathscr{E}$ and which covers the latter and plays the same role in the category of differential equations that the tangent bundle plays in the category of smooth manifolds. It is used to construct recursion operators for symmetries of $\mathscr{E}$ and symplectic structures on $\mathscr{E}$. In this chapter we give the solution to Problem 1.18 and also prepare a basis to solution of Problems 1.20 (Chap. 7) and 1.22 (Chap. 8).

6.1 General Theory

Let $\mathscr{E} \subset J^\infty(n, m)$ be an equation given as zeros of the vector function $F = (F^1, \dots, F^r) \in \mathscr{F}(n, m; r) = P$. Consider the equation $\mathscr{T}\mathscr{E} \subset J^\infty(n, 2m)$ given by the system

$$\begin{aligned} &F^j(x, \dots, u_\sigma, \dots) = 0, \\ &\ell_{F^j}(x, \dots, u_\sigma, \dots, q_\sigma, \dots) = 0, \end{aligned} \qquad j = 1, \dots, r,$$

where $q = (q^1, \dots, q^m)$ is a new auxiliary dependent variable and the projection $\mathrm{t} = \mathrm{t}_{\mathscr{E}} : \mathscr{T}\mathscr{E} \to \mathscr{E}$, $(x, u_\sigma, q_\sigma) \mapsto (x, u_\sigma)$. This projection is a covering called the *tangent covering* to $\mathscr{E}$.

Remark 6.1 By a number of reasons (see below), it is convenient to treat the fibers variables as odd. Thus, $\mathscr{T}\mathscr{E}$ becomes a supermanifold, [84], see also [74, 82].

Example 6.1 The tangent covering to the Burgers equation is

$$\begin{aligned} u_t &= uu_x + u_{xx}, \\ q_t &= u_x q + uq_x + q_{xx}. \end{aligned}$$

Thus, the covering equation is of evolutionary form too, and the functions x, t, u_k, and q_k, where the subscript denotes the kth x-derivative, $k = 0, 1, \dots$, may be taken for the internal coordinates on $\mathscr{T}\mathscr{E}$. Then the total derivatives on $\mathscr{T}\mathscr{E}$ acquire the form

J. Krasil'shchik et al., *The Symbolic Computation of Integrability Structures for Partial Differential Equations*, Texts and Monographs in Symbolic Computation,
https://doi.org/10.1007/978-3-319-71655-8_6

$$\tilde{D}_x = D_x + \sum_{k\geq 0} q_{k+1}\frac{\partial}{\partial q_k}, \quad \tilde{D}_t = D_t + \sum_{k\geq 0} \tilde{D}_x^k\,(u_1q_0 + u_0q_1 + q_2)\,\frac{\partial}{\partial q_k},$$

where D_x and D_t are the total derivatives on $\mathscr{E}$ (see Remark 2.3). It is straightforward to check that $[\tilde{D}_x, \tilde{D}_t] = 0$ and thus t is a covering indeed.

Example 6.2 Consider the Gibbons-Tsarev equation $u_{yy} + u_xu_{xy} - u_yu_{xx} + 1 = 0$. Its tangent covering is described by the system

$$\begin{aligned} &u_{yy} + u_xu_{xy} - u_yu_{xx} + 1 = 0,\\ &q_{yy} + u_xq_{xy} - u_yq_{xx} + u_{xy}q_x - u_{xx}q_y = 0. \end{aligned}$$

If the equation is given in the two-component form then $\mathscr{T}\mathscr{E}$ is

$$\begin{aligned} &v_y + wv_x = \frac{1}{w-v}, && w_y + vw_x = \frac{1}{v-w},\\ &q_y + wq_x + v_xr = \frac{q_x - r_x}{(w-v)^2}, && r_y + vr_x + w_xq = \frac{r_x - q_x}{(w-v)^2}, \end{aligned}$$

where q, r are nonlocal variables in the tangent covering.

Example 6.3 The tangent covering to the universal hierarchy equation $u_{yy} = u_tu_{xy} - u_yu_{xt}$ is

$$\begin{aligned} u_{yy} &= u_tu_{xy} - u_yu_{xt},\\ q_{yy} &= u_tq_{xy} - u_yq_{xt} + u_{xy}q_t - u_{xt}q_y. \end{aligned}$$

Let us now give a coordinate-free definition of the tangent covering. To this end, we must define functions on $\mathscr{T}\mathscr{E}$ and the Cartan connection in the bundle $\mathscr{T}\mathscr{E} \to M$.

Consider an equation $\mathscr{E} \subset J^\infty(\pi)$, and let us say that a differential form $\omega \in \Lambda^1(\mathscr{E})$ is a *Cartan form* (or *vertical form*) if it vanishes on the Cartan distribution, i.e., if $i_Z(\omega) = 0$ for any $Z \in \mathscr{C}$. Denote by $\Lambda_v^1(\mathscr{E})$ the space of these forms. Then one has the decomposition

$$\Lambda^1(\mathscr{E}) = \Lambda_h^1(\mathscr{E}) + \Lambda_v^1(\mathscr{E}),$$

where $\Lambda_h^1(\mathscr{E})$ is the space of horizontal 1-forms (see Sect. 1.1). Let us also introduce the notation

$$\Lambda_v^p(\mathscr{E}) = \underbrace{\Lambda_v^1(\mathscr{E}) \wedge \dots \wedge \Lambda_v^1(\mathscr{E})}_{p \text{ times}}.$$

Then

$$\Lambda^i(\mathscr{E}) = \bigoplus_{p+q=i} \Lambda_h^q(\mathscr{E}) \otimes \Lambda_v^p(\mathscr{E}).$$

Define the *Cartan* (or *vertical*) *differential* by $d_v = d - d_h$, where d_h is the horizontal de Rham differential. Then

$$d_v\colon \Lambda_h^q(\mathscr{E}) \otimes \Lambda_v^p(\mathscr{E}) \to \Lambda_h^q(\mathscr{E}) \otimes \Lambda_v^{p+1}(\mathscr{E})$$

and

$$d_v \circ d_v = 0, \qquad [d_v, d_h] = 0,$$

where $[d_v, d_h] = d_v \circ d_h + d_h \circ d_v$ is the *anticommutator* of the Cartan and the horizontal de Rham differentials.

If (x, u_σ^j) are internal coordinates on $\mathscr{E}$ and $f = f(x, u_\sigma^j)$ is a smooth function the action of d_v is completely determined by the formulas

$$d_v(f) = \sum_{j,\sigma} \frac{\partial f}{\partial u_\sigma^j} \omega_\sigma^j,$$

where the forms $\omega_\sigma^j = \left(du_\sigma^j - \sum_i u_{\sigma i}^j\, dx^i\right)\Big|_{\mathscr{E}}$ constitute a basis in $\Lambda_v^1(\mathscr{E})$.

Consider now the vertical subbundle $t^v\colon T^v\mathscr{E} \to \mathscr{E}$ of its tangent bundle. This is a vector bundle, and we can identify fiber-wise linear functions on $T^v\mathscr{E}$ with Cartan forms on $\mathscr{E}$ by setting

$$\omega(\tilde{\theta}) = i_v(\omega), \qquad \tilde{\theta} \in T^v\mathscr{E}, \tag{6.1}$$

where $\tilde{\theta} = (\theta, v)$, $\theta \in \mathscr{E}$, and v is a vertical tangent vector at the point θ.

Let us now prolong the Cartan connection $\mathscr{C}\colon D(M) \to D(\mathscr{E})$ to a connection $\tilde{\mathscr{C}}\colon D(M) \to D(T^v\mathscr{E})$. To this end, consider a vector field $X \in D(M)$ and set

$$\tilde{\mathscr{C}}_X(\omega) = L_{\mathscr{C}_X}(\omega) = d(i_{\mathscr{C}_X}\omega) + i_{\mathscr{C}_X}(d\omega).$$

Then we have:

- The equality $\tilde{\mathscr{C}}_X(f\omega) = \mathscr{C}_X(f)\omega + f\tilde{\mathscr{C}}_X(\omega)$ is valid for any $f \in \mathscr{F}(\mathscr{E})$, i.e., $\tilde{\mathscr{C}}$ is a prolongation of $\mathscr{C}$.
- The correspondence $X \mapsto \tilde{\mathscr{C}}_X$ is $C^\infty(M)$-linear. Indeed, due to (6.1)

$$\tilde{\mathscr{C}}_{fX}(\omega) = df \wedge i_{\mathscr{C}_X}(\omega) + f\tilde{\mathscr{C}}_X(\omega) = f\tilde{\mathscr{C}}_X(\omega),$$

  because ω is a Cartan form and the field $\mathscr{C}_X$ lies in the Cartan distribution. Hence, the correspondence $X \mapsto \tilde{\mathscr{C}}_X$ is a connection.
- This connection is flat, because $L_{[X,Y]} = [L_X, L_Y]$.

Consequently, the bundle $t^v\colon T^v\mathscr{E} \to \mathscr{E}$ is a covering. This covering is equivalent to the tangent one.

Remark 6.2 There exists a canonical identification between the vertical tangent bundle t^v over $\mathscr{E}$ and the subbundle $\ker \ell_{\mathscr{E}}$ in the bundle of horizontal jets $\bar{\pi}^h_\infty : J^\infty_h(\bar{\pi}) \to \mathscr{E}$, where, as before, $\bar{\pi}$ denotes the pullback $\pi^*_\infty(\pi)$.

Recall now that we consider the fibers of the tangent covering to be odd, and using the above arguments, let us define the algebra of superfunctions on $\mathscr{T}\mathscr{E}$ as $\mathscr{F}^s(\mathscr{T}\mathscr{E}) = \Lambda^*_v(\mathscr{E})$, where $\Lambda^*_v(\mathscr{E})$ is the exterior algebra:

$$\Lambda^*_v(\mathscr{E}) = \bigoplus_{p \geq 0} \Lambda^p_v(\mathscr{E}).$$

Understood in such a way, the tangent covering possesses the following important features:

- The Cartan differential $d_v : \mathscr{F}^s(\mathscr{T}\mathscr{E}) \to \mathscr{F}^s(\mathscr{T}\mathscr{E})$ becomes an odd nilpotent vector field on the supermanifold $\mathscr{T}\mathscr{E}$. In examples below this field is denoted by $\mathbf{X}$.
- Any shadow in the tangent covering can be canonically lifted to a nonlocal symmetry in this covering. Indeed, consider a shadow, i.e., a vertical derivation $S : \mathscr{F}(\mathscr{E}) \to \mathscr{F}^s(\mathscr{T}\mathscr{E})$ that satisfies the equality

$$\tilde{\mathscr{C}}_X \circ S = S \circ \mathscr{C}_X$$

for any $X \in D(M)$. Locally, any such a derivation is of the form

$$S = \mathbf{E}_\omega = \sum_{j,\sigma} D_\sigma(\omega^j) \frac{\partial}{\partial u^j_\sigma},$$

where $\omega = (\omega^1, \dots, \omega^m)$ and ω^j are Cartan forms. Define the *inner product*

$$i_S(\rho) = \sum_{j,\sigma} D_\sigma(\omega^j) \wedge i_{\partial/\partial u^j_\sigma}(\rho), \qquad \rho \in \Lambda^*_v(\mathscr{E}),$$

and "vertical" Lie derivative

$$\begin{aligned} L^v_S(\rho) &= [d_v, i_S](\rho) \\ &= \sum_{j,\sigma} \left(d_v \left(D_\sigma(\omega^j) i_{\partial/\partial u^j_\sigma}(\rho) \right) - (-1)^{|\omega^j| \cdot |\rho|} \left(D_\sigma(\omega^j) i_{\partial/\partial u^j_\sigma} d_v(\rho) \right) \right) \end{aligned} \tag{6.2}$$

where $|\cdot|$ is the degree and $[\cdot, \cdot]$ denotes the *graded commutator*. The vector field $\tilde{S} = L^v_S$ is the desired lift of the shadow S.

Another two properties of the covering $\mathrm{t} : \mathscr{T}\mathscr{E} \to \mathscr{E}$ are also essential for the theory and applications:

(1) there is a one-to-one correspondence between the maps $\varphi\colon \mathscr{E} \to \mathscr{T}\mathscr{E}$ whose differential takes the total derivatives on $\mathscr{E}$ to the hull of those on $\mathscr{T}\mathscr{E}$ (i.e., preserve the Cartan distributions) and symmetries of $\mathscr{E}$;
(2) there is a one-to-one correspondence between cosymmetries of $\mathscr{E}$ and conservation laws on $\mathscr{T}\mathscr{E}$ linear in q_σ. Conservation laws on $\mathscr{T}\mathscr{E}$ obtained in such a way are called *canonical*.

Remark 6.3 Property 2 is a consequence of the Green formula (2.7). Namely, let ψ be a cosymmetry of $\mathscr{E}$ and $\bar{\psi}$ be its extension to the space $J^\infty(n, 2m)$. Then the identity $\langle \ell_{\mathscr{E}}(q), \psi\rangle - \langle q, \ell^*_{\mathscr{E}}(\psi)\rangle = 0$ on $\mathscr{T}\mathscr{E}$ leads to

$$\langle \ell_F(q), \psi\rangle - \langle q, \ell^*_F(\psi)\rangle = d_h\bar{\omega}(\psi),$$

on the ambient jet space, where $\bar{\omega}(\psi)$ is an $(n-1)$-horizontal form that depends on ψ. Its restriction $\omega(\psi) = \bar{\omega}(\psi)|_{\mathscr{T}\mathscr{E}}$ is closed and $\psi \mapsto \omega(\psi)$ is the desired correspondence.

In the case of evolution equations $u^j_t = f(x, t, u, \dots, u^\alpha_\sigma, \dots)$ in two independent variables x and t, the x-component of $\omega(\psi)$ is $\sum_{j=1}^m \psi^j q^j$.

Example 6.4 Consider the Burgers equation and its cosymmetry $\psi = 1$. Then we obtain the conservation law

$$\omega = q\,dx + (uq + q_x)\,dt$$

on $\mathscr{T}\mathscr{E}$.

Example 6.5 Consider the cosymmetries of the Gibbons-Tsarev equation from Example 4.2. Then to $\psi_0 = 1$ there corresponds the conservation law

$$\omega_0 = (q_y + 2u_x q_x)\,dx + (u_y q_x + u_x q_y)\,dy$$

on $\mathscr{T}\mathscr{E}$, to $\psi_1/2 = u_x$ we put into correspondence

$$\omega_1 = (3u_x^2 q_x + 2u_x q_y + 2u_y q_x)\,dx + (2u_x u_y q_x + u_x^2 q_y + 2u_y q_y - 2q)\,dy$$

and the conservation law that corresponds to $\psi_{-5} = u_{xxx}$ is $\omega_{-5} = X_{-5}\,dx + Y_{-5}\,dy$, where

$$X_{-5} = u_{xxx}q_y + u_y q_{xxx} - \frac{1}{2}u_{xx}^2 q_x - u_x u_{xx} q_{xx},$$

$$\begin{aligned} Y_{-5} &= u_{xxy}q_y + u_y q_{xxy} + \frac{1}{2}u_{xx}^2 q_y + u_y u_{xx} q_{xx} \\ &\quad - u_{xx}u_{xy}q_x - u_x u_{xy} q_{xx} - u_x u_{xx} q_{xy} - u_{xy} q_{xy} - q_{xx}. \end{aligned}$$

Example 6.6 If we take the cosymmetries

$$\psi_1(T) = \frac{T}{u_y^2}, \qquad \psi_1(X) = X$$

of the universal hierarchy equation from Example 4.3 (recall that $T = T(t)$ and $X = X(x)$), then the corresponding conservation laws on $\mathcal{TE}$ will be

$$\omega_1(T) = \frac{T}{u_y^2}\big((u_y q_t - u_t q_y)\, dy \wedge dt - q_y\, dx \wedge dt\big)$$

and

$$\omega_1(X) = X\big((q_y + uq_{xt} + u_{xt}q)\, dx \wedge dy + (u_{xy}q + uq_{xy})\, dx \wedge dt\big),$$

respectively.

Remark 6.4 In the case when a cosymmetry ψ is a generating function of a conservation law $\omega = \sum_i X_i\, dx^1 \wedge \dots \wedge \widehat{dx^i} \wedge \dots \wedge dx^n$ on $\mathcal{E}$, the corresponding conservation law on $\mathcal{TE}$ is

$$\tilde{\omega} = \sum_i \ell_{X_i}(q)\, dx^1 \wedge \dots \wedge \widehat{dx^i} \wedge \dots \wedge dx^n$$

i.e., the linearization of ω.

Perhaps, the most important property of the tangent covering follows from the fact that it coincides with the Δ-covering when $\Delta = \ell_{\mathcal{E}}$. Indeed, let us apply Proposition 1.2 to our case. Namely, assume that $\mathcal{E}$ is of the form $\mathcal{E} = \{F = 0\}$, $F = (F^1, \dots, F^r) \in P = \Gamma(\bar{\xi})$ (recall that "bar" denotes the pullback of a bundle over M to $\mathcal{E}$ by $\pi_\infty \colon \mathcal{E} \to M$), and consider two spaces $R = \Gamma(\bar{\eta})$, $R' = \Gamma(\bar{\eta}')$, $\operatorname{rank}(\eta) = l$, $\operatorname{rank}(\eta') = l'$. Let

$$\Upsilon = (\Upsilon^1, \dots, \Upsilon^l), \qquad \Upsilon^\alpha = \sum_{j,\sigma} \upsilon^\alpha_{j,\sigma} q^j_\sigma,$$

be a fiber-wise linear vector function on $\mathcal{TE}$ and $\nabla = \nabla_\Upsilon \colon \varkappa \to R$ be a matrix $\mathcal{C}$-differential operator of the form

$$\nabla = \begin{pmatrix} \sum_\sigma \upsilon^1_{1,\sigma} D_\sigma & \dots & \sum_\sigma \upsilon^j_{1,\sigma} D_\sigma \\ \dots & \dots & \dots \\ \sum_\sigma \upsilon^1_{l,\sigma} D_\sigma & \dots & \sum_\sigma \upsilon^j_{l,\sigma} D_\sigma \end{pmatrix}.$$

Then the following result is valid:

Proposition 6.1 *The diagram*

$$\begin{array}{ccc} \varkappa & \xrightarrow{\ \ell_{\mathcal{E}}\ } & P \\ {\scriptstyle\nabla}\downarrow & & \downarrow{\scriptstyle\nabla'} \\ R & \xrightarrow{\ \square\ } & R' \end{array}$$

is commutative for some $\mathscr{C}$-differential operator $\nabla' : P \to R'$ if and only if

$$\widetilde{\square}(\Upsilon) = 0, \tag{6.3}$$

where $\widetilde{\square}$ is the natural lift of $\square$ to the tangent covering. Vice versa, to any such an operator $\square$, one can put into correspondence a fiber-wise linear function $\Upsilon_\square$ that satisfies Eq. (6.3), and this function vanishes if and only if ∇ is of the form $\nabla = \nabla'' \circ \ell_{\mathscr{E}}$ for some $\mathscr{C}$-differential operator $\nabla'' : P \to R$.

In particular, the operators ∇ obtained by the above-described procedure take elements of $\ker \ell_{\mathscr{E}} = \operatorname{sym}(\mathscr{E})$ to those of $\ker \square$.

The construction of Proposition 6.1 can be reformulated in the following geometrical way. Let $\mathscr{E}_\square \subset J_h^\infty(\bar{\xi})$ be the *$\mathscr{C}$-differential equation* defined by the operator $\square$, and let us preserve the notation $\bar{\xi}_\infty^h$ for the restriction of the projection $\bar{\xi}_\infty^h : J_h^\infty(\bar{\xi}) \to \mathscr{E}$ to $\mathscr{E}_\square$. Then the diagram

$$\begin{array}{ccc} \mathscr{T}\mathscr{E} & \xrightarrow{\ \Upsilon_\nabla\ } & \mathscr{E}_\square \\ & {\scriptstyle \mathtt{t}}\searrow \quad \swarrow{\scriptstyle \bar{\xi}_\infty^h} & \\ & \mathscr{E}, & \end{array}$$

where $\Upsilon_\nabla([s]_\theta^\infty) = [\nabla(s)]_\theta^\infty$, $\theta \in \mathscr{E}$, is commutative. If $s \in \Gamma(\tau)$ preserves the Cartan distributions (i.e., is a symmetry of $\mathscr{E}$), then $\Phi_\nabla \circ s$ is a section of $\bar{\xi}_\infty^h$ that preserves the Cartan distributions as well, i.e., it is a solution of $\mathscr{E}$. The correspondence between symmetries of $\mathscr{E}$ given by this construction is local, i.e., is expressed in terms of $\mathscr{C}$-differential operators. But, as we shall see in the next chapters, one often needs nonlocal objects in important applications.

To this end, consider the canonical conservation law ω on $\mathscr{T}\mathscr{E}$ associated to a cosymmetry $\psi \in \operatorname{cosym}(\mathscr{E})$, and assume that this is a two-component conservation law, i.e.,

$$\omega = X_1\, dx^1 \wedge dx^3 \wedge \cdots \wedge dx^n + X_2\, dx^2 \wedge dx^3 \wedge \cdots \wedge dx^n,$$

where

$$X_i = \sum_{i,j,\sigma} X_{i,j,\sigma} q_\sigma^j, \qquad X_{i,j,\sigma} \in \mathscr{F}(\mathscr{E}), \quad i = 1, 2,$$

and q_σ^j are internal coordinates on $\mathscr{T}\mathscr{E}$. Then, as it was indicated in Sect. 3.1.2, we can construct the covering τ_ω with nonlocal variables w^ρ that are defined by the relations

$$w_{x^1}^\rho = D_\rho(X_1), \qquad w_{x^2}^\rho = D_\rho(X_2), \qquad w_{x^i}^\rho = w^{\rho i}, \quad i > 2, \tag{6.4}$$

where ρ is a multi-index consisting of the integers $3, \dots, n$.

Assume now that a vector function

$$\Upsilon = (\Upsilon^1, \dots, \Upsilon^l), \qquad l = \operatorname{rank}(\xi),$$

where

$$\Upsilon^\alpha = \sum_{j,\sigma} \upsilon^\alpha_{j,\sigma} q^j_\sigma + \sum_\rho \upsilon^\alpha_\rho w^\rho, \qquad \upsilon^\alpha_{j,\sigma}, \upsilon^\alpha_\rho \in \mathscr{F}(\mathscr{E}),$$

is a solution of the equation

$$\widetilde{\square}(\Upsilon) = 0, \tag{6.5}$$

where $\widetilde{\square}$ is the lift of $\square$ to τ_ω. Any such a solution defines a covering

$$\tau_\Upsilon \colon W \to \mathscr{E}_\square$$

and we obtain the following commutative diagram

$$\begin{array}{ccccc} & & W & & \\ & \swarrow^{\tau_\omega} & & \searrow^{\tau_\Upsilon} & \\ \mathscr{T}\mathscr{E} & & & & \mathscr{E}_\square \\ & \searrow_{\mathrm{t}} & & \swarrow_{\bar{\xi}^h_\infty} & \\ & & \mathscr{E}. & & \end{array}$$

In other words, any fiber-wise linear solution to Eq. (6.5) gives rise to a Bäcklund transformation between $\mathscr{T}\mathscr{E}$ and $\mathscr{E}_\square$ and the correspondence between symmetries of $\mathscr{E}$ and solutions of $\mathscr{E}_\square$ becomes nonlocal.

Example 6.7 Consider the conservation law ω on the tangent covering to the Burgers equation (Example 6.4). The corresponding covering $\tau_\omega \colon W \to \mathscr{T}\mathscr{E}$ is defined by the equalities

$$w_x = q, \qquad w_t = uq + q_x.$$

Then to a solution $\Upsilon = \sum_k \upsilon^k q_k + \upsilon w$ of Eq. (6.5), there corresponds the operator

$$\nabla_\Upsilon = \underbrace{\sum_k \upsilon^k D_x^k}_{\text{local part}} + \upsilon D_x^{-1}.$$

Example 6.8 Let $\mathscr{E}$ be the Gibbons-Tsarev equation, and consider the conservation law ω_0 from Example 6.5. The corresponding covering is

$$w_x = q_y + 2u_x q_x, \qquad w_y = u_y q_x + u_x q_y.$$

The corresponding nonlocal operators will take the form

$$\text{local part} + \upsilon D_x^{-1} \circ (D_y + 2u_x D_x)$$

or

$$\text{local part} + \upsilon D_y^{-1} \circ (u_y D_x + u_x D_y).$$

Example 6.9 Consider finally the conservation law $\omega_1(T)$ of the UHE presented in Example 6.9. Then the corresponding nonlocal variable satisfies the equations

$$w_y = \frac{T}{u_y^2}(u_y q_t - u_t q_y), \qquad w_x = -\frac{T}{u_y^2} q_y$$

and leads to the operators of the form

$$\text{local part} + \upsilon D_y^{-1} \circ \left(\frac{1}{u_y} D_t - \frac{u_t}{u_y^2} D_y \right)$$

or

$$\text{local part} - \upsilon D_x^{-1} \circ \frac{T}{u_y^2} D_y.$$

On the other hand, the conservation law $\omega_1(X)$ gives the covering

$$w_y = X(u_{xt} q + q_y + u q_{xt}), \qquad w_t = u_{xy} q + u q_{xy}$$

to which the operators

$$\text{local part} + \upsilon D_y^{-1} \circ (u_{xt} + D_y + u D_x D_t)$$

or

$$\text{local part} + \upsilon D_t^{-1} \circ (u_{xy} + u D_x D_y)$$

correspond.

6.2 Examples

Let us now discuss the CDE implementation of the above constructions.

6.2.1 Korteweg-de Vries Equation

The defining equation for the tangent covering are

$$u_t = uu_x + u_{xxx},$$

$$q_t = u_x q + uq_x + q_{xxx}.$$

As we already saw, the linearization in CDE is already computed using odd variables, and the second equation of the above system is obtained during that computation. The system can be loaded in CDE as follows (example file kdv_tcl1.red):

```
% the jet space
indep_var:={x,t};
dep_var:={u};
odd_var:={q};
total_order:=10;
% the system
principal_der:={u_t};
de:={u*u_x+u_3x};
principal_odd:={q_t};
de_odd:={u_x*q+u*q_x+q_3x};
% the call
cde({indep_var,dep_var,odd_var,total_order},
   {principal_der,de,principal_odd,de_odd});
```

The equations are treated on a different footing, as we are dealing with the even variables x, t, and u and the odd variable q. As we are dealing with a supermanifold, we stress that odd variables may only appear as odd polynomials; an odd monomial is expressed in the form `ext(p_x,p_t2x,...)`, or just `p_x,...` if the monomial is linear (see also Sect. 2.2.2).

After calling CDE, the system is ready to compute total derivatives of any expression in parametric even and odd coordinates. Note that total derivatives are always automatically lifted to the tangent covering by CDE if odd variables and the corresponding odd equation are present in the CDE jet space.

The conservation law on $\mathcal{T}\mathcal{E}$ that corresponds to the cosymmetry $\psi_1 = 1$ (see Sect. 4) is $\omega_1 = q\,dx + (uq + q_2)\,dt$. To the cosymmetry $\psi_2/2 = u$, there corresponds the conservation law $\omega_2 = uq\,dx + (uq_2 + u^2 q - u_1 q_1 + u_2 q)\,dt$. The above conservation laws can be found as follows. After loading the tangent covering as above, add the following weights:

```
deg_indep_var:={-1,-3};
deg_dep_var:={2};
deg_odd_var:={1};
```

After having created weighted monomials as in the previous examples:

```
l_grad_even:=der_deg_ordering(0,all_parametric_der);
gradmon:=graded_mon(1,10,l_grad_even);
gradmon:={1} . gradmon;
```

it is necessary to create weighted monomials which are linear in odd variables. Since the two conservation laws have degrees of the coefficients of *dx*, respectively, equal to 1 and 3, the weight equation resulting from the conservation law Eq. (3.1) yields that the coefficients of *dt* must have degrees equal to 3 and 5, respectively. After having set up the list of odd variables according to their weights

```
l_grad_odd:=
  {1} . der_deg_ordering(1,all_parametric_odd);
```

the following commands create all monomials which have the previous weights from the lists of weighted even variables `gradmon` and odd variables `l_grad_odd`

```
linodd3:=mkalllinodd(gradmon,l_grad_odd,3,3);
linodd5:=mkalllinodd(gradmon,l_grad_odd,5,5);
```

Then we define the equation

```
c1x:=q; % degree 1
c2x:=u*q; % degree 3

c1t:=for each el in linodd3 sum
  c(ctel:=ctel+1)*el; % degree 3
c2t:=for each el in linodd5 sum
  c(ctel:=ctel+1)*el; % degree 5

equ 1:=td(c1t,x)-td(c1x,t);
equ 2:=td(c2t,x)-td(c2x,t);
```

and solve it in the usual way to find the above conservation laws on the tangent covering:

```
c1t;

q*u + q_2x;

c2t;

q*u**2  + q*u_2x + q_2x*u - q_x*u_x;
```

6.2.2 Dispersionless Boussinesq System

The tangent covering is

$$
\begin{aligned}
u_t &= ww_x + v_x, & q^u_t &= q^w w_x + wq^w_x + q^v_x,\\
v_t &= -uw_x - 3wu_x, & q^v_t &= -q^u w_x - uq^w_x - 3(q^w u_x + wq^u_x)\\
w_t &= u_x, & q^w_t &= q^u_x;
\end{aligned}
$$

in this case, q^u, q^v, and q^w are "the tangent variables" that correspond to u, v, and w. The conservation law that corresponds to a cosymmetry $\psi = (\psi^u, \psi^v, \psi^w)$ is computed by the formula

$$\omega_\psi = \left(q^u\psi^u + q^v\psi^v + q^w\psi^w\right)dx + \left(q^u(\psi^w - 3w\psi^v) + q^v\psi^u + q^w(w\psi^u - u\psi^v)\right)dt.$$

We make use of the above when computing the recursion operator for the dispersionless Boussinesq equation in Chap. 7. We will describe the initial part of the corresponding program file bou_ro1.red. We initialize the jet space as follows:

```
indep_var:={x,t};
dep_var:={u,v,w};
odd_var:={p,q,r,o1,o2,o3};
total_order:=8;
```

The notation `o1,o2,` and `o3` stands for the three odd nonlocal variables corresponding to the conservation laws ω_1, ω_2, and ω_3 in Sect. 3.2.2 (see Sect. 4.2.2 for the corresponding cosymmetries). The even equations are

```
principal_der:={u_t,v_t,w_t};
de:={w*w_x + v_x, - u*w_x - 3w*u_x,u_x};
```

and the odd equations, including the nonlocal variables, are

```
principal_odd:={p_t,q_t,r_t,
  o1_t,o1_x,
  o2_t,o2_x,
  o3_t,o3_x};
de_odd:={q_x + r*w_x + r_x*w,
  - (p*w_x + 3*p_x*w + 3*r*u_x + r_x*u),p_x,
  p,r,
  q+r*w,p,
  - w*p - u*r,q + 2*w*r};
```

6.2.3 *Camassa-Holm Equation*

The tangent covering of the Camassa-Holm equation is

$$\alpha(u_t + 3uu_x) - u_{txx} = 2u_x u_{xx} + uu_{xxx},$$
$$\alpha(q_t + 3qu_x + 3uq_x) - q_{xxx} = 2q_x u_{xx} + 2u_x q_{xx} + qu_{xxx} + uq_{xxx}.$$

The conservation laws on $\mathcal{TE}$ that correspond to cosymmetries ψ_{ω_1} and ψ_{ω_2} (see Sect. 4.2.3) are

$$\omega_1 = \alpha q_0^0\,dx + \left(q_1^1 + q_0^2 u_0^0 + q_0^1 u_0^1 + q_0^0(u_0^2 - 3\alpha u_0^0)\right)dt,$$
$$\omega_2 = (\alpha u_0^0 - u_0^2)q_0^0\,dx + \left(u_0^0 q_1^1 - q_1^0 u_0^1 + q_0^2 (u_0^0)^2 + q_0^0 u_0^0(2u_0^2 - 3\alpha u_0^0)\right)dt.$$

We will describe the initial part of the program file ch_ro1.red, where we generate the tangent covering together with two nonlocal variables with the code:

```
indep_var:={t,x};
dep_var:={u};
odd_var:={q,o1,o2};
total_order:=8;
principal_der:={u_3x};
de:={(alpha*(u_t + 3*u*u_x) - u_t2x - 2*u_x*u_2x)/u};

% The nonlocal odd variables
o1x:=alpha*q;
o1t:=(u_2x - 3*u*alpha)*q + u_x*q_x +u*q_2x +q_tx;

o2x:=(alpha*u - u_2x)*q;
o2t:=((u_2x - 3*u*alpha)*u + u*u_2x)*q
  + (u_x*u - u*u_x)*q_x +u**2*q_2x - u_x*q_t +u*q_tx;

principal_odd:={q_3x,o1_x,o1_t,o2_x,o2_t};
de_odd:={( - alpha*q*u_t + alpha*q_t*u
  + 3*alpha*q_x*u**2 + 2*q*u_2x*u_x + q*u_t2x
  - 2*q_2x*u*u_x - q_t2x*u - 2*q_x*u*u_2x)/u**2,
  o1x,o1t,o2x,o2t};
```

It is worth to verify that the odd variables are well-defined:

```
td(s1,t,x) - td(s1,x,t);
```

and the result is zero (the same holds for `s2`).

Weights are consistently defined by

```
deg_indep_var:={-2,-1};
deg_dep_var:={1};
deg_odd_var:={1,2,3};
```

6.2.4 Multi-dimensional Examples

Consider two multi-dimensional examples.

6.2.4.1 The Kadomtsev-Petviashvili Equation

The tangent covering for this equation is

$$u_{yy} = u_{tx} - u_x^2 - uu_x - \frac{1}{12}u_{xxxx}$$

$$q_{yy} = q_{tx} - 2u_x q_x - u_x x q - u q_{xx} - \frac{1}{12}q_{xxxx}$$

and can be generated in CDE for the scalar presentation (2.14) as follows (initial part of the program file kp_tan1.red):

```
indep_var:={t,x,y};
dep_var:={u};
odd_var:={q};
total_order:=6;
principal_der:={u_2y};
de:={u_tx-u_x**2-u*u_2x-(1/12)*u_4x};
principal_odd:={q_2y};
de_odd:={
( - 12*q*u_2x - 12*q_2x*u - q_4x + 12*q_tx
  - 24*q_x*u_x)/12
};
```

and consistent weights are given by

```
deg_indep_var:={-3,-1,-2};
deg_dep_var:={2};
deg_odd_var:={1};
```

The tangent covering for the evolutionary presentation (2.15) is generated as follows (initial part of the program file kpev_tan1.red)

```
indep_var:={t,x,y};
dep_var:={u,v};
odd_var:={p,q};
total_order:=6;
principal_der:={u_y,v_y};
de:={v_x,u_t - u*u_x - (1/12)*u_3x};
principal_odd:={p_y,q_y};
de_odd:={q_x,( - 12*p*u_x - p_3x + 12*p_t
  - 12*p_x*u)/12};
```

and consistent weights are

```
deg_indep_var:={-3,-1,-2};
deg_dep_var:={2,3};
deg_odd_var:={1,2};
```

6.2.4.2 The Plebanski Equation

The equation is

$$F = u_{tt}u_{xx} - u_{tx}^2 + u_{xz} + u_{ty} = 0; \tag{6.6}$$

it is Lagrangian (Sect. 2.3.4). This means that its linearization is self-adjoint, $\ell_F = \ell_F^*$, so that the tangent and cotangent (see Sect. 9) coverings coincide, its odd equation being

$$\ell_F(p) = p_{xz} + p_{ty} - 2u_{tx}p_{tx} + u_{2x}p_{2t} + u_{2t}p_{2x} = 0. \tag{6.7}$$

It is not difficult to realize that the above equation can be written in explicit conservative form as

$$\begin{aligned} p_{xz} + p_{ty} + u_{tt}p_{xx} + u_{xx}p_{tt} - 2u_{tx}p_{tx} \\ = D_x(p_z + u_{tt}p_x - u_{tx}p_t) + D_t(p_y + u_{xx}p_t - u_{tx}p_x) = 0, \end{aligned}$$

thus the corresponding conservation law is

$$\upsilon(1) = (p_y + u_{xx}p_t - u_{tx}p_x)\, dx \wedge dy \wedge dz + (u_{tx}p_t - p_z - u_{tt}p_x)\, dt \wedge dy \wedge dz. \tag{6.8}$$

The above conservation law has *two* components. For this reason we can introduce a potential variable r. Namely, we can assume that

$$r_x = p_y + u_{xx}p_t - u_{tx}p_x, \quad r_t = u_{tx}p_t - p_z - u_{tt}p_x. \tag{6.9}$$

This is a new nonlocal variable for the (co)tangent covering of the Plebanski equation. We can load the Plebanski equation together with its nonlocal variable r as follows (program file ple_nlv1.red):

```
indep_var:={t,x,y,z};
dep_var:={u};
odd_var:={p,r};
deg_indep_var:={-1,-1,-4,-4};
deg_dep_var:={1};
deg_odd_var:={1,4};
total_order:=6;
```

```
principal_der:={u_xz};
de:={-u_ty+u_tx**2-u_2t*u_2x};
% rhs of the equations that define
% the nonlocal variable
rt:= - p_z - u_2t*p_x + u_tx*p_t;
rx:= p_y + u_2x*p_t - u_tx*p_x;
% We add conservation laws as new nonlocal
% odd variables;
principal_odd:={p_xz,r_x,r_t};
%
de_odd:={-p_ty+2*u_tx*p_tx-u_2x*p_2t-u_2t*p_2x,rx,rt};
```

We can easily verify that the integrability condition for the new nonlocal variable holds:

```
td(r,t,x) - td(r,x,t);
```

the result is 0.

Chapter 7
Recursion Operators for Symmetries

Abstract A recursion operator for symmetries of an equation $\mathcal{E}$ is a $\mathcal{C}$-differential operator $\mathcal{R}: \varkappa = \mathcal{F}(\mathcal{E}; m) \to \varkappa$ that takes symmetries of $\mathcal{E}$ to themselves. We expose below a computational theory of such operators based on the tangent covering techniques. The simplest version of this theory relates to local operators, but in reality all recursion operators, except for the case of linear equations with constant coefficients, are nonlocal. Such operators, in general, act on shadows of symmetries only. Unfortunately, to the best of our knowledge, a self-contained theory for these operators (as well as for nonlocal operators of other types that are considered below) does not exist at the moment, but some reasonable ideas can be applied to particular classes of examples nevertheless. In this chapter, we give the solution to Problems 1.20 and 1.28.

7.1 General Theory

Consider an equation $\mathcal{E}$, and let $\mathcal{R} = (\mathcal{R}_i^j)$ be a matrix with the entries consisting of the $\mathcal{C}$-differential operators

$$\mathcal{R}_i{}^j = \sum_\sigma a^j_{i,\sigma} D_\sigma, \qquad i, j = 1, \dots, m,$$

on $\mathcal{E}$. Using the construction described in Sect. 6.1, consider the function $\varphi_{\mathcal{R}} = (\varphi^1_{\mathcal{R}}, \dots, \varphi^m_{\mathcal{R}})$, where

$$\varphi^j_{\mathcal{R}} = \sum_{i,\sigma} a^j_{i,\sigma} q^i_\sigma,$$

with q^i_σ being local coordinates in the fibers of the tangent covering of $\mathcal{E}$. Recall that $\varphi_{\mathcal{R}}$ is an odd vector function on $\mathcal{T}\mathcal{E}$.

J. Krasil'shchik et al., *The Symbolic Computation of Integrability Structures for Partial Differential Equations*, Texts and Monographs in Symbolic Computation,
https://doi.org/10.1007/978-3-319-71655-8_7

Denote by $\tilde{\ell}_{\mathcal{E}}$ the lift of the linearization operator $\ell_{\mathcal{E}}$ to $\mathcal{T}\mathcal{E}$, and consider the equation

$$\tilde{\ell}_{\mathcal{E}}\left(\varphi_{\mathcal{R}}\right) = 0. \tag{7.1}$$

Recall that the tangent covering to $\mathcal{E}$ is identified with the Δ-covering with $\Delta = \ell_{\mathcal{E}}$. Consequently, due to Proposition 1.2, the following fact, essential for computations, holds: solutions of the above equation are in one-to-one correspondence with the classes of operators

$$\frac{\{\mathcal{R} \mid \ell_{\mathcal{E}} \circ \mathcal{R} = \Delta \circ \ell_{\mathcal{E}}\}}{\{\mathcal{R} \mid \mathcal{R} = \nabla \circ \ell_{\mathcal{E}}\}},$$

where $\Delta\colon \mathcal{F}(\mathcal{E};r) \to \mathcal{R}(\mathcal{E};r)$ and $\nabla\colon \mathcal{F}(\mathcal{E};r) \to \mathcal{F}(\mathcal{E};m)$ are some $\mathcal{C}$-differential operators. In other words, non-zero solutions of (7.1) provide us with nontrivial recursion operators of the equation $\mathcal{E}$. Note also that functions $\varphi_{\mathcal{E}}$ that satisfy Eq. (7.1) are shadows of symmetries in the tangent covering.

Consider a tutorial example.

Example 7.1 Let

$$\mathcal{E}\colon\ u_t = u_{xx}$$

be the one-dimensional heat equation. The space $\mathcal{T}\mathcal{E}$ is obtained by adding the equation $q_t = q_{xx}$. Taking, as before, the functions x, t, $u_k = \partial^k u/\partial x^k$ for internal coordinates on $\mathcal{E}$ and adding the odd coordinates $q_k = \partial^k q/\partial x^k$ in the fibers of the tangent covering, we obtain the total derivatives on $\mathcal{T}\mathcal{E}$ in the form

$$\tilde{D}_x = \frac{\partial}{\partial x} + \sum_{k\ge 0}\left(u_{k+1}\frac{\partial}{\partial u_k} + q_{k+1}\frac{\partial}{\partial q_k}\right), \quad \tilde{D}_t = \frac{\partial}{\partial t} + \sum_{k\ge 0}\left(u_{k+2}\frac{\partial}{\partial u_k} + q_{k+2}\frac{\partial}{\partial q_k}\right).$$

Solving the equation

$$\tilde{\ell}_{\mathcal{E}}(\Phi) \equiv \tilde{D}_t(\Phi) - \tilde{D}_x^2(\Phi) = 0,$$

where $\Phi = a_0 q + a_1 q_1 + \cdots + a_k q_k$, for $k \le 1$, we obtain the lowest order solutions q, q_1, and $tq_1 + x/2$, which correspond to the operators

$$\mathcal{R}_0 = \mathrm{id}, \qquad \mathcal{R}_{10} = D_x, \qquad \mathcal{R}_{11} = tD_x + \frac{x}{2},$$

and it can be shown that the entire algebra of recursion operators is multiplicatively generated by these ones. Since $[\mathcal{R}_{10}, \mathcal{R}_{11}] = 1/2$, this algebra is isomorphic to the universal enveloping algebra of the *three*-dimensional *Heisenberg algebra*.

However, the existence of a local recursion operator is an exception, and in the majority of interesting examples, they are nonlocal. A standard way to enrich the initial settings with nonlocal variables is to use cosymmetries of the equation $\mathcal{E}$ and construct the corresponding conservation laws on $\mathcal{T}\mathcal{E}$ (see Sect. 6.1).

Example 7.2 The tangent covering to the Burgers equation $u_t = uu_x + u_{xx}$ is given by the additional equation

$$q_t = u_x q + u q_x + q_{xx}$$

Let us choose internal coordinates x, t, u_k, q_k in the standard way and consider the total derivatives

$$\tilde{D}_x = \frac{\partial}{\partial x} + \sum_{k\geq 0}\left(u_{k+1}\frac{\partial}{\partial u_k} + q_{k+1}\frac{\partial}{\partial q_k}\right),$$

$$\tilde{D}_t = \frac{\partial}{\partial t} + \sum_{k\geq 0}\left(\tilde{D}_x^k(u_0u_1 + u_2)\frac{\partial}{\partial u_k} + \tilde{D}_x^k(u_1q_0 + u_0q_1 + q_2)\frac{\partial}{\partial q_k}\right).$$

The defining equation for recursion operator is

$$\tilde{D}_t(\Phi) = u_1\Phi + u_0\tilde{D}_x(\Phi) + \tilde{D}_x^2(\Phi), \tag{7.2}$$

where $\Phi = a_0q_0 + \dots + a_kq_k$ and the coefficients a_i are functions on $\mathcal{E}$. It is an easy exercise to prove that the only solution of this form is αq_0, $\alpha \in \mathbb{R}$. Thus, the only *local* recursion operator for symmetries of the Burgers equation is proportional to the identical one.

To obtain a nontrivial result, consider the only cosymmetry $\psi = 1$ of the Burgers equation and construct the corresponding conservation law

$$\omega = q_0\,dx + (uq_0 + q_1)\,dt$$

on $\mathcal{T}\mathcal{E}$ (cf. Example 6.7) and construct the one-dimensional covering $\tau\colon \widetilde{\mathcal{T}\mathcal{E}} \to \mathcal{T}\mathcal{E}$ defined by

$$w_x = q_0, \qquad w_t = u_0q_0 + q_1,$$

where w is an odd nonlocal variable. The total derivatives take the form

$$\tilde{\tilde{D}}_x + \frac{\partial}{\partial w}, \qquad \tilde{\tilde{D}}_t + (uq_0 + q_1)\frac{\partial}{\partial w}$$

on this covering. Consider the linearization operator lifted to τ, and denote it by $\tilde{\tilde{\ell}}_{\mathcal{E}}$. The equation

$$\tilde{\tilde{\ell}}_{\mathcal{E}}(\Phi) = 0$$

has a two-dimensional space of solutions of the form $\Phi = bw + a_0q_0 + a_1q_1 + a_2q_2$. This space is spanned by the functions

$$\Phi_0 = q_0, \quad \Phi_2 = q_2 + \frac{1}{2}u_0q_0 + \frac{1}{2}u_1w,$$

while the corresponding recursion operators are

$$\mathscr{R}_0 = \mathrm{id}, \quad \mathscr{R}_2 = D_x^2 + \frac{1}{2}u_0 + \frac{1}{2}u_1D_x^{-1},$$

which are, of course, well known.

Example 7.3 For the sine-Gordon equation

$$u_{xy} = \sin(u)$$

the tangent covering is described by the system

$$u_{xy} = \sin(u),$$
$$q_{xy} = \cos(u)q.$$

The equation admits the recursion operator[1]

$$\mathscr{R} = D_x^2 + u_x^2 - u_xD_x^{-1} \circ u_{xx},$$

e.g., see details in [24].

Let us generalize the above example to the case of an arbitrary evolution equation $\mathscr{E}$ in two independent variables,

$$u_t{}^j = f^j(x, t, u, \dots, u_k, \dots), \qquad u = (u^1, \dots, u^m), \quad j = 1, \dots, m.$$

Assume that $\psi_l = (\psi_l^1, \dots, \psi_l^m)$ are cosymmetries of this equation. Then the corresponding nonlocal variables enjoy the relations

$$w_{l,x} = \psi_l^1 q^1 + \dots + \psi_l^m q^m. \tag{7.3}$$

Consider the corresponding covering $\tau\colon \widetilde{\mathscr{T}\mathscr{E}} \to \mathscr{T}\mathscr{E}$ and the lift $\tilde{\tilde{\ell}}_{\mathscr{E}}$ of the linearization operator to this covering. Now we look for solutions of the equation

$$\tilde{\tilde{\ell}}_{\mathscr{E}}(\Phi) = 0,$$

[1] This and many other examples that are discussed below were taken from [145, 146].

where

$$\Phi^j = \sum_{i,\sigma} a^j_{i,\sigma} q^i_\sigma + \sum_l b_l{}^j w_l, \qquad j = 1, \dots, m, \tag{7.4}$$

with the nonlocal variables w_l satisfying Eq. (7.3) for the chosen set of cosymmetries $\psi_l = (\psi_l^1, \dots, \psi_l^m)$. To any such a solution, there corresponds the matrix operator $\mathscr{R}_\Phi$ that acts on generating functions $\varphi = (\varphi^1, \dots, \varphi^m)$ of symmetries and whose action is described by the formulas

$$(\mathscr{R}_\Phi(\varphi))^j = \sum_{i,\sigma} a^j_{i,\sigma} D_\sigma(\varphi^i) + \sum_l b_l^j D_x^{-1}\left(\sum_i \psi_l^i \varphi^i\right), \tag{7.5}$$

for all $i = 1, \dots, m$. Similarly to the local variables q^j_σ, the nonlocal ones are also considered to be odd. Operators of this form are called *weakly nonlocal*, cf. [89]. Note that the vector functions $b_l = (b_l^1, \dots, b_l^m)$ that appear in the representation (7.5) are symmetries of $\mathscr{E}$.

Remark 7.1 Nonlocal recursion operators of the form (7.5) (as well as more general ones), generally speaking, take symmetries to nonlocal shadows of symmetries. This happens, for example, if one applies the recursion operator of the KdV equation (see Sect. 7.2.1 below) to the Galilean boost. The problem of locality of these shadows is analyzed in [130], where a sufficient condition involves the hereditary property of the operator, see below. However, there are examples of non-hereditary operators which produce a hierarchy of local symmetries [129].

7.1.1 Variational Nijenhuis Bracket

Let $\mathscr{R}$ be a recursion operator for symmetries of some equation $\mathscr{E}$. Its *Nijenhuis torsion* is given by the formula

$$N_{\mathscr{R}}(\varphi_1, \varphi_2) = \{R(\varphi_1), R(\varphi_2)\} - R\big(\{R(\varphi_1), \varphi_2\} + \{\varphi_1, R(\varphi_2)\} - R\{\varphi_1, \varphi_2\}\big), \tag{7.6}$$

where $\varphi_1, \varphi_2 \in \operatorname{sym}(\mathscr{E})$. This definition is inconvenient, because one needs to know particular symmetries φ_1 and φ_2. We shall give an alternative (and more general) one in terms of fiber-wise linear functions on $\mathscr{T}\mathscr{E}$ that correspond to the operators at hand.

Let us begin with the local case and assume that $\Phi = (\Phi^1, \dots, \Phi^m)$ is a q-linear vector functions on $\mathscr{T}\mathscr{E}$: $\Phi^j = \sum_{i,\sigma} a^j_{i,\sigma} q^i_\sigma$, where $a^j_{i,\sigma}$ are functions on $\mathscr{E}$. Introduce the odd derivations

$$\mathbf{E}_\Phi = \sum_{j,\sigma} D_\sigma(\Phi^j) \frac{\partial}{\partial u^j_\sigma}.$$

Let also $\mathbf{X}$ be the odd vector field on $\mathcal{TE}$ defined by

$$\mathbf{X}(u^j_\sigma) = q^j_\sigma, \qquad \mathbf{X}(x^i) = \mathbf{X}(q^j_\sigma) = 0. \tag{7.7}$$

Remark 7.2 Note that $\mathbf{X}$ is an evolutionary vector field on $\mathcal{TE}$ with the generating section $q = (q^1, \dots, q^m)$ and it commutes with the total derivatives. This field is always a symmetry of the tangent equation. Recall also (see Chap. 6) that $\mathbf{X}$ actually is the Cartan differential understood as a vector field on $\mathcal{TE}$.

Then for another function $\bar{\Phi} = (\bar{\Phi}^1, \dots, \bar{\Phi}^m)$ of the same type, we set

$$\mathbf{E}_\Phi(\bar{\Phi})^j = \sum_{i,\sigma} \left(\mathbf{E}_\Phi(\bar{a}^j_{i,\sigma}) q^i_\sigma - \bar{a}^j_{i,\sigma} D_\sigma \mathbf{X}(\Phi^i) \right) \tag{7.8}$$

and

$$\{\!\{\Phi, \bar{\Phi}\}\!\} = \mathbf{E}_\Phi(\bar{\Phi}) + \mathbf{E}_{\bar{\Phi}}(\Phi), \tag{7.9}$$

The expression (7.9) is called the (variational) *Nijenhuis bracket* of Φ and $\bar{\Phi}$. It is even (quadratic) function on $\mathcal{TE}$.

Let us discuss the nonlocal case now and confine ourselves with the equations $\mathcal{E}$ in two independent variables x and y. Consider cosymmetries $\psi_1, \dots, \psi_k$ of $\mathcal{E}$, where

$$\psi_i = (\psi_i^1, \dots, \psi_i^m),$$

and let $w_1, \dots, w_k$ be the nonlocal variables on $\mathcal{TE}$ associated to these cosymmetries and satisfying the defining equations

$$w_{i,x} = \psi_i^1 q^1 + \dots + \psi_i^m q^m, \qquad i = 1, \dots, k.$$

By definition, the functions w_i are odd. Take two functions $\Phi = (\Phi^1, \dots, \Phi^m)$ and $\bar{\Phi} = (\bar{\Phi}^1, \dots, \bar{\Phi}^m)$ linear in q and w:

$$\Phi^j = \sum_{i,\sigma} a^j_{i,\sigma} q^i_\sigma + \sum_{l=1}^k b^j_l w_l,$$

where $a_{i,\sigma}$ and b^j_l are smooth functions on $\mathcal{E}$. Let us set

$$\mathbf{E}_\Phi(\bar{\Phi}^j) = \sum_{i,\sigma} \left(\mathbf{E}_\Phi(\bar{a}^j_{i,\sigma}) q^i_\sigma - \bar{a}^j_{i,\sigma} D_\sigma(\mathbf{X}\Phi^i) \right)$$
$$+ \sum_l \left(\mathbf{E}_\Phi(\bar{b}^j_l) w_l - \bar{b}^j_l D_x^{-1} \left(\mathbf{X} \sum_i \bar{\psi}^i_l \Phi^i \right) \right) \tag{7.10}$$

and define the nonlocal Nijenhuis bracket using (7.10) and Eq. (7.9). Of course, the main problem lies in computation of term containing the nonlocal variables and the operator D_x^{-1}. We cannot formulate a general theory at the moment, but shall demonstrate reasonable attitudes in the examples below.

7.1.2 Hereditary Operators

A recursion operator $\mathscr{R} = \mathscr{R}_\Phi$ is said to be *hereditary* if the function Φ enjoys the relation $\{\!\{\Phi, \Phi\}\!\} = 2\mathbf{E}_\Phi(\Phi) = 0$ (which is equivalent to vanishing of the Nijenhuis torsion (7.6)). Let φ be a symmetry of $\mathscr{E}$ and $\mathscr{R}$ be a hereditary recursion operator invariant with respect to φ. Then all the symmetries of the form $\mathscr{R}^l(\varphi)$, provided they are local, pair-wise commute [70].

Example 7.4 Let us prove that the operators $\mathscr{R}_{10}$ and $\mathscr{R}_{11}$ from Example 7.1 are hereditary. With the first one, it is obvious, because from formulas (7.8) to (7.9) it readily follows that any local operator with constant coefficients is hereditary. Consider the operator $\mathscr{R}_{11}$ now and the linear function

$$\Phi = tq_x + \frac{1}{2}xq$$

that corresponds to this operator. Then

$$\mathbf{E}_\Phi(\Phi) = \mathbf{E}_\Phi\left(tq_x + \frac{1}{2}xq\right) = -tD_x\mathbf{X}\left(tq_x + \frac{1}{2}xq\right) - \frac{1}{2}x\mathbf{X}\left(tq_x + \frac{1}{2}xq\right) = 0$$

due to the definition of $\mathbf{X}$.

Example 7.5 Consider the recursion operator

$$\mathscr{R} = D_x + \frac{1}{2}u + \frac{1}{2}u_xD_x^{-1}$$

of the Burgers equation. The corresponding function on $\mathscr{T}\mathscr{E}$ is

$$\Phi = q_x + \frac{1}{2}uq + \frac{1}{2}u_xw,$$

where the nonlocal variable w corresponds to the cosymmetry $\psi = 1$ and is defined by $w_x = q$. We need to compute the quantity

$$\begin{aligned}\mathbf{E}_\Phi(\Phi) &= \mathbf{E}_\Phi\left(q_x + \frac{1}{2}uq + \frac{1}{2}u_xw\right)\\ &= \mathbf{E}_\Phi(q_x) + \frac{1}{2}\mathbf{E}_\Phi(u)q + \frac{1}{2}u\mathbf{E}_\Phi(q) + \frac{1}{2}\mathbf{E}_\Phi(u_x)w + \frac{1}{2}u_x\mathbf{E}_\Phi(w).\end{aligned}$$

To proceed with computations, note that the action of D_x on fiber-wise linear functions has no kernel, from where it immediately follows that

$$\mathbf{X}(w) = 0, \qquad \mathbf{E}_\Phi(w) = -\frac{1}{2}qw.$$

Then

$$\mathbf{E}_\Phi(q_x) = -D_x\mathbf{X}(\Phi) = -\frac{1}{2}D_x(q \cdot q + q_x w) = -\frac{1}{2}(q_{xx}w + q_x q),$$

because $q \cdot q = 0$ due to oddness of q. Further,

$$\begin{aligned}
\mathbf{E}_\Phi(u)q &= \left(q_x + \frac{1}{2}uq + \frac{1}{2}u_x w\right)q = \left(q_x + \frac{1}{2}u_x w\right)q,\\
u\mathbf{E}_\Phi(q) &= -u\mathbf{X}\left(q_x + \frac{1}{2}uq + \frac{1}{2}u_x w\right) = -\frac{1}{2}uq_x w,\\
\mathbf{E}_\Phi(u_x)w &= D_x\left(q_x + \frac{1}{2}uq + \frac{1}{2}u_x w\right)w\\
&= \left(q_{xx} + \frac{1}{2}u_x q + \frac{1}{2}uq_x + \frac{1}{2}u_x x w + \frac{1}{2}u_x q\right)w\\
&= \left(q_{xx} + \frac{1}{2}u_x q + \frac{1}{2}uq_x + \frac{1}{2}u_x q\right)w.
\end{aligned}$$

Consequently, we obtain

$$\begin{aligned}
\mathbf{E}_\Phi(\Phi) = &-\frac{1}{2}(q_{xx}w + q_x q) + \frac{1}{2}\left(q_x + \frac{1}{2}u_x w\right)q - \frac{1}{4}uq_x w\\
&+ \frac{1}{2}\left(q_{xx} + \frac{1}{2}u_x q + \frac{1}{2}uq_x + \frac{1}{2}u_x q\right)w - \frac{1}{4}u_x qw\\
&= \frac{1}{4}u_x(wq + qw) = 0,
\end{aligned}$$

because the variables w and q anti-commute.

Example 7.6 Consider the recursion operator for symmetries of the KdV equation

$$\mathscr{R} = D_x^2 + \frac{2}{3}u + \frac{1}{3}u_x D_x^{-1}$$

(see Sect. 7.2.1) with the corresponding function

$$\Phi = q_{xx} + \frac{2}{3}uq + \frac{1}{3}u_x w,$$

where, as in the previous example, the nonlocal variable w corresponds to the cosymmetry $\psi = 1$. Here, as in the previous example, the total derivative operator D_x acts monomorphically on fiber-wise line functions and consequently

$$\mathbf{E}_\Phi(w) = -\frac{1}{3}qw.$$

Then, computations which are quite similar to those from Example 7.5 show that $\mathscr{R}$ is a hereditary operator, i.e., $\mathbf{E}_\Phi(\Phi) = 0$.

Example 7.7 The dispersionless Boussinesq system

$$\begin{aligned} u_t &= ww_x + v_x, \\ v_t &= -uw_x - 3wu_x, \\ w_t &= u_x \end{aligned}$$

admits two recursion operators $\mathscr{R}_1$ and $\mathscr{R}_2$ (see computations in Sect. 7.2.2), which satisfy the relation $[\mathscr{R}_2, \mathscr{R}_1] = \mathscr{R}_1^2$. Neither of them is hereditary, but their linear combination $\mathscr{R} = 3\mathscr{R}_1 + 4\mathscr{R}_2$ is; see [61].

7.1.3 Recursion Operators as Bäcklund Transformations

Presentation of recursion operators as pseudo-differential ones is intuitively convenient but lacks of geometrical clarity and becomes somewhat inappropriate in the multi-dimensional situation. A purely geometrical approach was proposed in [96] and consists of treating recursion operators as auto-Bäcklund transformations

$$\begin{array}{ccc} & \widetilde{\mathscr{T}\mathscr{E}} & \\ {}^{\tau_1}\swarrow & & \searrow^{\tau_2} \\ \mathscr{T}\mathscr{E} & & \mathscr{T}\mathscr{E} \end{array} \tag{7.11}$$

of the tangent equation. In the simplest cases, similar to the ones considered above, such an interpretation looks as follows.

Example 7.8 Consider the recursion operator for the Burgers equation (see Example 7.2). Denote by q^1 and q^2 the fiber coordinates in two copies of $\mathscr{T}\mathscr{E}$. Then

$$\begin{aligned} w_x &= q^1, \\ w_t &= q^1_x + uq^2, \\ q^2 &= q^1_x + \frac{1}{2}uq^1 + \frac{1}{2}u_x w \end{aligned}$$

is the Bäcklund transformation corresponding to this operator.

In multi-dimensional cases, the diagram (7.11) becomes more sophisticated.

Example 7.9 Consider the universal hierarchy equation

$$u_{yy} = u_t u_{xy} - u_y u_{tx}.$$

Then

$$\begin{aligned} u_y q_x^1 &= -q_y^2 + u_{xy} q^1, \\ u_y q_y^1 &= u_y q_t^2 - u_t q_y^2 - (u_y u_{tx} - u_t u_{xy}) q^1 \end{aligned}$$

is a recursion operator for this equation. Moreover, the Bäcklund transformation

$$\begin{aligned} q_t^2 &= q_y^1 - u_t q_x^1 + u_{tx} q^1, \\ q_y^2 &= -u_y q_x^1 + u_{xy} q^1 \end{aligned}$$

is also a recursion operator, which is natural to consider as the inverse to the first one, see [103].

7.2 Examples

Let us pass now to discussion of the computer realization.

7.2.1 *Korteweg-de Vries Equation*

Using the cosymmetry $\psi = 1$ of the KdV equation, introduce the nonlocal variable w on $\mathscr{T}\mathscr{E}$ defined by $w_x = 1 \cdot q = q$. Let $\Phi = a_2 q_{xx} + a_1 q_x + a_0 q + bw$. Solving the equation $\tilde{\ell}_{\mathscr{E}}(\Phi) = 0$, one finds two independent solutions: $\Phi = q$ (the identical operator) and

$$\Phi = q_{xx} + \frac{2}{3} u q + \frac{1}{3} u_x w,$$

to which the classical Lenard recursion operator

$$\mathscr{R} = D_x^2 + \frac{2}{3} u + \frac{1}{3} u_x D_x^{-1}$$

corresponds.

In CDE one should load the equation together with the nonlocal variable. The equations for recursion operators can be solved in two ways, namely, by the weight approach and CRACK. Let us begin with weights, program file kdv_ro1.red:

```
principal_der:={u_t};
de:={u*u_x+u_3x};
principal_odd:={q_t,r_x,r_t};
de_odd:={u_x*q+u*q_x+q_3x,q,u*q+q_2x};
```

The weights are assigned as

```
deg_indep_var:={-1,-3};
deg_dep_var:={2};
deg_odd_var:={2,1};
cde_grading(deg_indep_var,deg_dep_var,deg_odd_var);
```

Then we shall load the linearization as a $\mathscr{C}$-differential operator lifted to the tangent covering. Total derivatives are always automatically lifted to the tangent covering by CDE if odd variables are present in CDE's jet space, even if linearization is defined exactly in the same way as in the purely even case:

```
mk_cdiffop(ell_kdv,1,{1},1);
for all sym let ell_kdv(1,1,sym)
 = td(sym,t)-u*td(sym,x)-u_x*sym-td(sym,x,3);
```

Then, we define the ansatz. Since $\mathscr{C}$-differential operators are linear operators, the ansatz must be linear in odd variables; in the weight approach, the ansatz must be a weighted polynomial. Hence, we define the ansatz sym as a weighted polynomial which is linear in odd variables and whose monomials have weights ranging from 1 to 6:

```
l_grad_mon:=der_deg_ordering(0,all_parametric_der);
l_grad_odd:=der_deg_ordering(1,all_parametric_odd);
gradmon:=graded_mon(1,10,l_grad_mon);
gradmon:={1} . gradmon;
l_grad_odd:={1} . l_grad_odd;
linodd:=mkalllinodd(gradmon,l_grad_odd,1,6);
sym:=(for each el in linodd sum
  (c(ctel:=ctel+1)*el));
```

After solving in the usual way, we obtain

```
sym;

2*c(5)*q*u + 3*c(5)*q_2x + c(5)*r*u_x + c(2)*q
```

where the constant c(2) multiplies the trivial recursion operator (it is just the identity operator) and the constant c(5) multiplies the Lenard recursion operator.

We could proceed to solving an overdetermined system of linear PDEs using CRACK. To this aim, we load the equation, its tangent covering, and the nonlocal variable as before. Then we define a linear function of odd variables with coefficients that depend on a prescribed set of even variables in an arbitrary way. We are describing the program file kdv_ro2.red. We generate an ansatz containing even and odd variables up to the order 2:

```
ansatz_evenvars:=for i:=0:2 join
  selectvars(0,i,dep_var,all_parametric_der);
ansatz_oddvars:=for i:=0:2 join
  selectvars(1,i,odd_var,all_parametric_odd);
```

Next code collects all unknown coefficients of the linear function of odd variables and defines the linear function itself:

```
coeff_oddvars:={};
sym:=0;
for each el in ansatz_oddvars do
  <<
    coeff_oddvars:=mkid(c_,el) . coeff_oddvars;
    sym:=mkid(c_,el)*el + sym;
  >>;
```

Then, we declare that the coefficient depend on even dependent variables of order not greater than 2:

```
for each el in ansatz_evenvars do
  for each ell in coeff_oddvars do
    depend ell,el;
```

We generate the equation

```
lin_eq:={ell_kdv(1,1,sym)};
```

Here the unknowns are the coefficients of odd variables:

```
unk:=coeff_oddvars;
```

Before starting CRACK, we split the equation collecting coefficients of monomials of variables from which the unknown functions are independent:

```
split_vars:=
  cde_difflist(all_parametric_der,ansatz_evenvars);
lin_eq_splitext:=splitext_list(lin_eq);
system_eq:=
  splitvars_list(lin_eq_splitext,split_vars);
```

The command `splitext_list` splits a list of equations into the list of the coefficients of all `ext` monomials. The call to CRACK:

```
load_package crack;
crack_sol:=crack(system_eq,{},unk,{});
```

The solutions are those described above.

7.2.2 Dispersionless Boussinesq System

We start the computation by loading the tangent covering with the corresponding odd variables, see Sect. 4.2.1. The program file is bou_ro1.red. We shall also define the weights of the variables:

```
deg_indep_var:={-1,-2};
deg_dep_var:={3,4,2};
deg_odd_var:={2,3,1,1,2,3};
```

After the call to CDE and the weights function:

```
cde_grading(deg_indep_var,deg_dep_var,deg_odd_var);
```

we compute all weighted monomials in order to produce an ansatz for the recursion operator. To this aim we first generate two lists of even and odd variables ordered by their weight:

```
gradlist:=der_deg_ordering(0,all_parametric_der);
gradlist_odd:=
  {1} . der_deg_ordering(1,all_parametric_odd);
```

Then we generate the list of lists of weighted monomials of even variables having weights from 1 to 10 (the figures depend on the range of investigation that we decide to consider):

```
gradmon:=graded_mon(1,10,l_grad_der);
gradmon:={1} . gradmon;
```

Finally, we generate a list of weighted monomials which are linear in odd variables and have weights from 1 to 8:

```
linodd:=mkalllinodd(gradmon,l_grad_odd,1,8);
```

We initialize a counter for the constants in the ansatz and the vector of equations:

```
ctel:=0;
operator c,equ;
```

After loading the linearization from Sect. 2.3.2, we construct the ansatz for the components phi(i) of the recursion operator:

```
operator phi;
for i:=1:nc do phi(i):=
  for each el in linodd sum c(ctel:=ctel+1)*el;
```

(here nc:=length(dep_var)), then we define the equations

```
for i:=1:nc do
  equ(i):=for j:=1:nc sum lbou(i,j,phi(j));
```

The equations are polynomials in all variables, the unknowns being the constants c(i). This means that we must split it with respect to odd and even variables, as follows:

```
tel:=splitext_opequ(equ,1,3);
tel2:=splitvars_opequ(equ,4,tel,vars);
```

In the above command `splitext_opequ`, the range `1-3` indicates the equations to be split with respect to odd variables. The resulting equations, which will be polynomial in even variables, will be arranged in the operator `equ` starting from the entry number `4` to the entry `tel`. The subsequent command `splitvars_opequ` makes only sense if the coefficients of the odd variables are all polynomials; it splits the coefficients from the previous step, generating `tel2 - tel` equations which will be purely linear algebraic equations in the constants `c(i)`. They are solved by the commands

```
put_equations_used tel2;
for i:=tel+1:tel2 do integrate_equation i;
```

We obtain two nontrivial recursion operators with the following expressions:

```
phi(1) := c(91)*o1*u*w_x + c(91)*o1*u_x*w
  - c(91)*o2*v_x - c(91)*o2*w*w_x - c(91)*o3*u_x
  - c(44)*o2*v_x - c(44)*o2*w*w_x
  - 2*c(44)*o3*u_x + 4*c(44)*p*v
  + 2*c(44)*p*w**2 + 3*c(44)*q*u
  + c(44)*r*u*w + c(3)*p;

phi(2) := c(91)*o1*u*u_x + 2*c(91)*o1*v_x*w
  + 3*c(91)*o1*w**2*w_x + c(91)*o2*u*w_x
  + 3*c(91)*o2*u_x*w - c(91)*o3*v_x
  + c(44)*o2*u*w_x + 3*c(44)*o2*u_x*w
  - 2*c(44)*o3*v_x - 11*c(44)*p*u*w
  + 4*c(44)*q*v - 3*c(44)*r*u**2
  - 6*c(44)*r*w**3 + c(3)*q;

phi(3) :=  - c(91)*o1*v_x - 2*c(91)*o1*w*w_x
  - c(91)*o2*u_x - c(91)*o3*w_x - c(44)*o2*u_x
  - 2*c(44)*o3*w_x + 3*c(44)*p*u + 2*c(44)*q*w
  + 4*c(44)*r*v + 8*c(44)*r*w**2 + c(3)*r;
```

It should be noted that a `CRACK` approach is much slower than the above approach.

7.2.3 Camassa-Holm Equation

We shall load the tangent covering with the two nonlocal odd variables as in Sect. 6.2.3, together with their weights (program file ch_ro1.red). After this, and after having loaded the linearization as a $\mathscr{C}$-differential operator, we start to search recursion operators. After having generated weighted monomials which are linear in odd variables as usual (here note that we shall add `alpha` as we did in Sect. 4.2.3), and after generating an ansatz `phi`, we define the equation to be solved:

```
equ 1:=num lch(1,1,phi);
```

Note that we just need that the numerator be zero. Solving the algebraic system coming from splitting the equation, we get

```
phi := (4*c(6)*alpha*q*u**2 - c(6)*o1*u_t
 + c(6)*o2*u_x - 2*c(6)*q*u*u_2x + c(6)*q_2t
 - c(6)*q_2x*u**2 - c(6)*q_x*u*u_x + c(6)*q_x*u_t
 + 4*c(5)*alpha**2*q + 4*c(2)*alpha*q*u
 + 2*c(2)*o1*u_x - 2*c(2)*q*u_2x - 2*c(2)*q_2x*u
 - 2*c(2)*q_tx - 2*c(2)*q_x*u_x
 + 4*c(1)*alpha*q)/(4*alpha);
```

which consist of one trivial and two nontrivial and nonlocal recursion operators, respectively, the coefficients of `c(1)` and `c(2)`, `c(6)`.

7.2.4 Heat Equation

It is simple to prove that the recursion operators for the heat equation are hereditary (see Sect. 7.4). We describe the computation in the program file h_nb1.red. We load the jet space and the equations:

```
indep_var:={t,x};
dep_var:={u};
odd_var:={q};
total_order:=10;
principal_der:={u_t};
de:={u_2x};
principal_odd:={q_t};
de_odd:={q_2x};
```

After calling `CDE`, we define the (local) recursion operator for the heat equation

```
mk_superfun(rh_sf,1,1);
rh_sf(1) := t*q_x + (1/2)*x*q,
```

and compute the Nijenhuis bracket of the operator with itself, using the formula (7.9);

```
nijenhuis_bracket(rh_sf,rh_sf,res_biv);
```

the result is

```
res_biv(1);
0;
```

Remark 7.3 At the moment, CDE can only compute the bracket of local recursion operators. There is no software procedure which computes the bracket of nonlocal recursion operators with sufficient generality. Concrete examples can be computed on a case-by-case basis, the problem being the simplification of expressions containing inverses of the total derivative operators. Of course, it is possible to introduce simplification rules; nonetheless, concrete experiments show that the simplification cannot be completed automatically in many nontrivial examples, thus requiring direct inspection.

As a future research topic we hope to obtain a systematic and general solution of the problem of computing the Nijenhuis bracket of nonlocal operators within our geometric framework.

7.2.5 Multi-dimensional Examples

Let us pass to multi-dimensional cases.

7.2.5.1 The Plebanski Equation

Let us consider the Plebanski equation. In Sect. 6.2.4 we found a nonlocal variable r for its tangent covering. Here we try to find a recursion operator using that nonlocal variable. The computations are done in the variables in which the Plebanski equation is presented (6.6), see [79]. We can load the jet space and the weights of the variables as (program file ple_ro1.red)

```
indep_var:={t,x,y,z};
dep_var:={u};
odd_var:={p,r};
deg_indep_var:={-1,-1,-4,-4};
deg_dep_var:={1};
deg_odd_var:={1,4};
total_order:=6;
```

and the equation of the tangent covering (together with the nonlocal variable) as

```
principal_der:={u_xz};
de:={-u_ty+u_tx**2-u_2t*u_2x};
rt:= - p_z - u_2t*p_x + u_tx*p_t;
rx:= p_y + u_2x*p_t - u_tx*p_x;
principal_odd:={p_xz,r_x,r_t};
de_odd:={-p_ty+2*u_tx*p_tx-u_2x*p_2t-u_2t*p_2x,
rx,rt};
```

We look for recursion operators which depend on r (which has weight 4); we produce the following ansatz for phi:

```
linodd:=mkalllinodd(gradmon,l_grad_odd,1,4);
phi:=for each el in linodd sum c(ctel:=ctel+1)*el);
```

then we solve the equation of shadows of symmetries:

```
equ 1:=td(phi,x,z)+td(phi,t,y)-2*u_tx*td(phi,t,x)
  +u_2x*td(phi,t,2)+u_2t*td(phi,x,2);
```

The solution is

```
phi := c(28)*r + c(1)*p
```

hence we obtain the identity operator p and the new nonlocal operator r (first obtained in [105]). The nonlocal variable r can be interpreted as a nonlocal symmetry of the Plebanski equation which belongs to the general class of the *partner symmetries*. Partner symmetries were first introduced in [90].

In the paper [105], the Plebanski equation is rewritten in coordinates that make it an evolution equation:

$$u_t = q, \qquad q_t = \frac{1}{u_{xx}}(q_x^2 - q_y - u_{xz}) \tag{7.12}$$

In these coordinates the nontrivial recursion operator has a much more complicated expression.

7.2.5.2 The rdDym Equation

In the paper [14], the rdDym equation is considered:

$$u_{ty} = u_x u_{xy} - u_y u_{xx}, \tag{7.13}$$

and two recursion operators are calculated: $\mathscr{R}_+$ and its inverse $\mathscr{R}_-$. The operator $\mathscr{R}_+$ was first found in [102] with methods developed in [99] (see also [132, 133]). Here we focus on the inverse recursion operator $\mathscr{R}_-$ and show that there is a simple derivation in our framework.

More precisely, from [14] we have the expression $\chi = \mathscr{R}_-(\hat{\chi})$, where

$$D_x(\chi) = D_t(\hat{\chi}) - u_x D_x(\hat{\chi}) + u_{xx}\hat{\chi}, \quad D_y(\chi) = -u_y D_x(\hat{\chi}) + u_{xy}\hat{\chi}. \tag{7.14}$$

It can be easily shown that the above recursion operator is a nonlocal variable on the cotangent covering which comes from a *two*-component conservation law ω of the form

$$\omega = (q_t - u_x q_x + u_{xx} q)dx \wedge dt + (-u_y q_x + u_{xy} q)dy \wedge dt. \tag{7.15}$$

Here we discuss the program file rddym_ro1.red. We load the jet space:

```
indep_var:={t,x,y};
dep_var:={u};
odd_var:={q,r};
total_order:=6;
```

then the even equation

```
principal_der:={u_ty};
de:={u_x*u_xy - u_y*u_2x};
```

and after defining the nonlocal variable

```
rx:=q_t - u_x*q_x + u_2x*q;
ry:= - u_y*q_x + u_xy*q;
```

we load the odd equation

```
principal_odd:={q_ty,r_x,r_y};
de_odd:={q_x*u_xy + u_x*q_xy - q_y*u_2x
  - u_y*q_2x,rx,ry};
```

After calling CDE, we define the linearization operator

```
mk_cdiffop(ell_rddym,1,{1},1);
for all psi let
  ell_rddym(1,1,psi) = td(psi,t,y) - u_xy*td(psi,x)
    - u_x*td(psi,x,y) + u_2x*td(psi,y)
    + u_y*td(psi,x,2);
```

The equation $\tilde{\ell}_{\mathscr{E}}(r) = 0$ is fulfilled (with $\mathscr{E}$ the rdDym equation):

```
ell_rddym(1,1,r);
0;
```

7.2.5.3 The Pavlov Equation

The Pavlov equation is also considered in [14]. Computations and considerations analogous to the rdDym equation apply in this case. More precisely, a recursion operator for the Pavlov equation was found in [102] with methods developed in [99] (see also [133]). Its inverse recursion operator admits a simple derivation in our framework. We describe the program file pav_ro1.red. The jet space is the same as the rdDym equation. We load the even equation

```
principal_der:={u_2y};
de:={u_tx + u_y*u_2x - u_x*u_xy};
```

and after defining the nonlocal variable

```
rx:=u_x*q_x + q_y - u_2x*q;
ry:=q_t + u_y*q_x - u_xy*q;
```

we load the odd equation

```
principal_odd:={q_2y,r_x,r_y};
de_odd:={q_2x*u_y  + q_tx - q_x*u_xy
  - q_xy*u_x + q_y*u_2x,rx,ry};
```

After calling CDE, we define the linearization operator

```
mk_superfun(lpav_sf,1,1);
lpav_sf(1):= - q_2x*u_y + q_2y - q_tx
  + q_x*u_xy + q_xy*u_x - q_y*u_2x;
conv_superfun2cdiff(lpav_sf,lpav);
```

The equation $\tilde{\ell}_{\mathscr{E}}(r) = 0$ is fulfilled (with $\mathscr{E}$ the Pavlov equation):

```
lpav(1,1,r);
0;
```

7.2.5.4 The Universal Hierarchy Equation

Finally, in [14] a recursion operator for the universal hierarchy equation is presented. The operator was found through Cartan's equivalence method in [103]. Computations and considerations analogous to the rdDym and the Pavlov equations apply in this case: the inverse of the above recursion operator can be obtained from a single nonlocal variable on the tangent covering. We describe the program file uh_ro1.red. The jet space is the same as the rdDym and Pavlov equations. We load the even equation

```
principal_der:={u_2y};
de:={u_t*u_xy - u_y*u_tx};
```

and after defining the nonlocal variable

```
rt:=q_y - u_t*q_x + u_tx*q;
ry:= - u_y*q_x + u_xy*q;
```

we load the odd equation

```
principal_odd:={q_2y,r_x,r_y};
de_odd:={q_2x*u_y  + q_tx - q_x*u_xy
  - q_xy*u_x + q_y*u_2x,rx,ry};
```

After calling CDE, we define the linearization operator

```
mk_cdiffop(ell_uh,1,{1},1);
for all psi let
ell_uh(1,1,psi) = td(psi,y,2) - u_t*td(psi,x,y)
  + u_y*td(psi,x,t) - u_xy*td(psi,t)
  + u_tx*td(psi,y);
```

The equation $\tilde{\ell}_{\mathscr{E}}(r) = 0$ is fulfilled (with $\mathscr{E}$ the universal hierarchy equation):

```
ell_uh(1,1,r);
0;
```

Remark 7.4 Inspired by the above examples (rdDym, Pavlov, and universal hierarchy equations), we might introduce a concept of partner symmetry that would fit the *three*-dimensional case. See [135] for a classification of scalar second-order *four*-dimensional PDEs admitting partner symmetries.

Chapter 8
Variational Symplectic Structures

Abstract A variational symplectic structure on an equation $\mathcal{E}$ is a $\mathcal{C}$-differential operator $\mathcal{S}\colon \varkappa = \mathcal{F}(\mathcal{E}; m) \to \hat{P} = \mathcal{F}(\mathcal{E}; r)$ that takes symmetries of $\mathcal{E}$ to cosymmetries and enjoys additional integrability properties. We expose here the computational theory of local symplectic structures and consider some instructive examples of nonlocal ones. In this chapter we give the solution to Problems 1.22, 1.23, and 1.28.

8.1 General Theory

Let a $\mathcal{C}$-differential operator $\mathcal{S}\colon \mathcal{F}(\mathcal{E}; m) = \varkappa \to \mathcal{F}(\mathcal{E}; r) = \hat{P}$, where $\mathcal{E} = \{F = 0\}$, $F \in P$, be of the form $\mathcal{S} = (\mathcal{S}_i^j)$ with

$$\mathcal{S}_i^j = \sum_\sigma a_{i,\sigma}^j D_\sigma, \qquad i = 1, \dots, m, \quad j = 1, \dots, r.$$

Using the construction of Sect. 6.1, consider the following vector function on $\mathcal{T}\mathcal{E}$:

$$\psi_{\mathcal{S}} = (\psi_{\mathcal{S}}^1, \dots, \psi_{\mathcal{S}}^r), \qquad \psi_{\mathcal{S}}^j = \sum_{i,\sigma} a_{i,\sigma}^j q_\sigma^i.$$

As before, $\psi_{\mathcal{S}}^j$ is an odd function.

Consider the lift $\tilde{\ell}_{\mathcal{E}}^*$ of the adjoint linearization operator $\ell_{\mathcal{E}}^*$ to the tangent equation $\mathcal{T}\mathcal{E}$ and the equation

$$\tilde{\ell}_{\mathcal{E}}^*(\psi_{\mathcal{S}}) = 0. \tag{8.1}$$

Then the solutions of (8.1) are in one-to-one correspondence with the classes of $\mathcal{C}$-differential operators

$$\frac{\{\, \mathcal{S} \mid \ell_{\mathcal{E}}^* \circ \mathcal{S} = \Delta \circ \ell_{\mathcal{E}} \,\}}{\{\, \mathcal{S} \mid \mathcal{S} = \nabla \circ \ell_{\mathcal{E}} \,\}},$$

J. Krasil'shchik et al., *The Symbolic Computation of Integrability Structures for Partial Differential Equations*, Texts and Monographs in Symbolic Computation,
https://doi.org/10.1007/978-3-319-71655-8_8

where $\Delta\colon \mathscr{F}(\mathscr{E};r) \to \mathscr{F}(\mathscr{E};m)$ and $\nabla\colon \mathscr{F}(\mathscr{E};r) \to \mathscr{F}(\mathscr{E};r)$ are some $\mathscr{C}$-differential operators. Any such an operator takes symmetries of the equation $\mathscr{E}$ to its cosymmetries.

Recall that we say that a conservation law $\omega \in \mathscr{L}(\mathscr{E})$ is *admissible* with respect to the operator $\mathscr{S}$ if $\psi_\omega = \mathscr{S}(\varphi)$ for some symmetry $\varphi \in \operatorname{sym}(\mathscr{E})$, where ψ_ω is the generating function of ω. For two admissible conservation laws ω and ω', we define their $\mathscr{S}$-bracket by

$$\{\omega, \omega'\}_{\mathscr{S}} = L_{\mathbf{E}_\varphi}(\omega'), \tag{8.2}$$

where L is the Lie derivative. Then $\mathscr{S}$ is a *variational symplectic structure* on $\mathscr{E}$ if the bracket (8.2):

(1) is skew-symmetric and
(2) satisfies the Jacobi identity,

i.e., is a *Poisson bracket*.

Conditions 1–2 are satisfied if, first,

$$\ell_{\mathscr{E}}^* \circ \mathscr{S} = \mathscr{S}^* \circ \ell_{\mathscr{E}}, \tag{8.3}$$

i.e., $\ell_{\mathscr{E}}^* \circ \mathscr{S}\colon \varkappa \to \hat{\varkappa}$ is a self-adjoint operator. Let $F \in \mathscr{F}(n,m;r)$ be a section that determines the equation $\mathscr{E}$ and $\bar{\mathscr{S}}\colon \mathscr{F}(n,m;m) \to \mathscr{F}(n,m;r)$ be an arbitrary extension of $\mathscr{S}$. Then due to the regularity condition of Sect. 2.1.1, there exists a $\mathscr{C}$-differential operator $\nabla\colon \mathscr{F}(n,m;r) \times \mathscr{F}(n,m;m) \to \mathscr{F}(n,m;m)$ such that

$$\ell_F^* \circ \bar{\mathscr{S}} - \bar{\mathscr{S}}^* \circ \ell_F = \nabla(F, \cdot).$$

Second, having the operator ∇ to satisfy the above conditions, one also needs the equality

$$\mathbf{E}_{\varphi_2}(\mathscr{S})(\varphi_1) - \mathbf{E}_{\varphi_1}(\mathscr{S})(\varphi_2) = \nabla|_{\mathscr{E}}^{*_1}(\varphi_1, \varphi_2) = 0,$$

where $\varphi_1, \varphi_2 \in \mathscr{F}(\mathscr{E};m)$ and the superscript ^{*_1} denotes conjugation with respect to the first argument of ∇.

In the case of evolution equation, Condition 1 amounts to skew-adjointness of $\mathscr{S}$, provided $\mathscr{S}$ corresponds to a solution of Eq. (8.1), while Condition 2 is checked in a simpler way than for general PDEs. Namely, consider the quadratic in the variables q_σ^j function

$$\Omega_{\mathscr{S}} = \sum_j q^j \psi_{\mathscr{S}}^i = \sum_j q^j \sum_{i,\sigma} a_{i,\sigma}^j q_\sigma^i.$$

Then $\mathscr{S}$ satisfies Condition 2 if and only if

$$\delta(\mathbf{X}(\Omega_{\mathscr{S}})) \equiv \frac{1}{2} \sum_i q^i \sum_\sigma (-1)^\sigma D_\sigma \frac{\partial \mathbf{X}(\Omega_{\mathscr{S}})}{\partial q_\sigma^i} = 0. \tag{8.4}$$

where $\mathbf{X}$ is the canonical vector field on $\mathscr{T}\mathscr{E}$ defined by (7.7).

Consider examples.

Example 8.1 For the beginning, take the heat equation $u_t = u_{xx}$. After simple computations, one immediately sees that this equation admits no nontrivial symplectic structure.

Example 8.2 Consider the potential KdV equation

$$u_t = 3u_x^2 + u_{xxx}.$$

Its linearization operator is

$$\ell_{\mathscr{E}} = D_t - 6u_x D_x - D_x^3$$

with the adjoint one equal to

$$\ell_{\mathscr{E}}^* = -D_t + 6D_x \circ u_x + D_x^3 = -D_t + 6u_{xx} + 6u_x D_x + D_x^3.$$

Consequently, the space $\mathscr{T}\mathscr{E}$ is given by the system

$$\begin{aligned} u_t &= 3u_x^2 + u_{xxx}, \\ q_t &= 6u_x q_x + q_{xxx}, \end{aligned} \tag{8.5}$$

while the defining equation for symplectic structures is

$$\tilde{D}_t(\Psi) = 6u_{xx}\Psi + 6u_x\tilde{D}_x(\Psi) + \tilde{D}_x^3(\Psi). \tag{8.6}$$

Here "tildes" denote the lift to $\mathscr{T}\mathscr{E}$, and

$$\Psi = a_0 q + a_1 q_1 + \cdots + a_k q_k$$

is a fiber-wise (odd) linear function on $\mathscr{T}\mathscr{E}$. Solving Eq. (8.6) on (8.5), one obtains the solution

$$\Psi = 2u_{xx}q + 4u_x q_x + q_{xxx}.$$

The corresponding operator $\mathscr{S}\colon \varkappa \to \hat{P}$ is of the form

$$\mathscr{S} = D_x^3 + 4u_x D_x + 2u_{xx} = D_x^3 + 2(u_x D_x + D_x \circ u_x).$$

This operator is obviously skew-adjoint. Let us check the condition (8.4) now. Keeping in mind that the variables q_k are odd, one has

$$\Omega_{\mathscr{S}} = q(2u_{xx}q + 4u_x q_x + q_{xxx}) = q(4u_x q_x + q_{xxx}).$$

Consequently,

$$\mathbf{X}(\Omega_{\mathscr{S}}) = 4q(\mathbf{X}(u_x)q_x) = 4qq_xq_x = 0,$$

and $\mathscr{S}$ is a symplectic structure.

Example 8.3 Consider potential Kaup-Kupershmidt equation

$$u_t = u_{xxxxx} + 10u_x u_{xxx} + \frac{15}{2}u_{xx}^2 + \frac{20}{3}u_x^3$$

and its tangent covering obtained by adding the equation

$$q_t = q_{xxxxx} + 10u_x q_{xxx} + 15u_{xx}q_{xx} + 10(u_{xxx} + 2u_x^2)q_x.$$

The defining equation for symplectic structures is

$$\begin{aligned}\tilde{D}_t(\Psi) &= (5u_{xxxx} + 40u_x u_{xx})\Psi \\ &\quad + (10u_{xxx} + 20u_x^2)\tilde{D}_x(\Psi) + 15u_{xx}\tilde{D}_x^2(\Psi) + 10u_x\tilde{D}_x^3(\Psi) + \tilde{D}_x^5(\Psi),\end{aligned}$$

where, as above, $\tilde{D}$ denotes the total derivative on $\mathscr{T}\mathscr{E}$ and $\Psi = \sum_i a_i q_i$, $a_i \in \mathscr{F}(\mathscr{E})$. The equation possesses a fifth-order solution

$$\Psi = q_5 + 10u_1 q_3 + 15u_2 q_2 + (9u_3 + 16u_1^2)q_1 + (2u_4 + 16u_1u_2)q$$

that corresponds to the operator

$$\begin{aligned}\mathscr{S} &= D_x^5 + 10u_1 D_x^3 + 15u_2 D_x^2 + (9u_3 + 16u_1^2)D_x^1 + (2u_4 + 16u_1u_2) \\ &= D_x^5 + 5(u_x D_x^3 + D_x^3 \circ u_x) - 3(u_{xxx}D_x + D_x \circ u_{xxx}) + 8(u_x^2 D_x D_x \circ u_x^2).\end{aligned}$$

which is skew-adjoint. Hence,

$$\begin{aligned}\Omega_{\mathscr{S}} = q\Psi &= q\big(q_5 + 10u_1 q_3 + 15u_2 q_2 + (9u_3 + 16u_1^2)q_1 + (2u_4 + 16u_1u_2)q\big) \\ &= q\big(q_5 + 10u_1 q_3 + 15u_2 q_2 + (9u_3 + 16u_1^2)q_1\big).\end{aligned}$$

Thus

$$\begin{aligned}\mathbf{X}(\Omega_{\mathscr{S}}) &= \mathbf{X}\big(q\big(q_5 + 10u_1 q_3 + 15u_2 q_2 + (9u_3 + 16u_1^2)q_1\big)\big) \\ &= q\big(10q_1q_3 + 15q_2q_2 + (9q_3 + 32q_1)q_1\big) \\ &= q\big(10q_1q_3 + 9q_3q_1\big) = qq_1q_3 = (qq_1q_2)_x,\end{aligned}$$

which means that $\delta(\mathbf{X}(\Omega_{\mathscr{S}})) = 0$, i.e, $\mathscr{S}$ is a symplectic structure for the potential Kaup-Kupershmidt equation.

Let us consider a nonlocal example now.

Example 8.4 Such an example is provided by the potential modified KdV equation. It reads

$$u_t = u_{xxx} + \frac{1}{3}u_x^3$$

Then the tangent covering is obtained by adding the equation

$$q_t = q_{xxx} + u_x^2 q_x,$$

and to find symplectic structures, we must solve the equation

$$\tilde{\ell}_{\mathscr{E}}^*(\Psi) = 2u_x u_{xx}\Psi + u_x^2\tilde{D}_x(\Psi) + \tilde{D}_x^3(\Psi) - \tilde{D}_t(\Psi) = 0 \tag{8.7}$$

on $\mathscr{T}\mathscr{E}$ for $\psi = a_0 q + \cdots + a_k q_k$. There is only the trivial solution in this setting.

Note that u_{xx} is a cosymmetry of the pmKdV equation, and introduce the nonlocal odd variable w by

$$w_x = u_{xx}q.$$

Then

$$D_x(\mathbf{X}(w)) = \mathbf{X}(w_x) = \mathbf{X}(u_{xx})q + u_{xx}\mathbf{X}(q) = q_{xx}q = D_x(q_x q)$$

and thus

$$\mathbf{X}w = q_x q$$

because, again, the action of D_x is injective on linear odd functions.

Let us solve Eq. (8.7) in the extended setting for $\psi = bw + a_0 q + \cdots + a_k q_k$. Then we immediately find the solution

$$\psi = q_{xxx} + \frac{2}{3}u_x^2 q_x + \frac{2}{3}u_x u_{xx} q - \frac{2}{3}u_{xx} w \tag{8.8}$$

with the corresponding operator

$$\mathscr{S} = D_x^3 + \frac{2}{3}u_x^2 D_x + \frac{2}{3}u_x u_{xx} - \frac{2}{3}u_{xx}D_x^{-1} = D_x^3 + \frac{2}{3}D_x u_x D_x^{-1} u_x D_x.$$

Evidently, $\mathscr{S}$ is skew-adjoint.

Let us generalize formula (8.4) to the nonlocal setting. Assume that an evolution equation $\mathscr{E} = \{u_t = f(x, t, u, \ldots, u_k)\}$ possesses a cosymmetry c, and consider the corresponding nonlocal variable w on $\mathscr{T}\mathscr{E}$ defined by $w_x = cq$. Suppose that $\psi = bw + a_0 q + \cdots + a_l q_l$ is a solution of the linearized equation on the tangent covering and the corresponding operator

$$\mathscr{S} = a_l D_x^l + \cdots + a_1 D_x + a_0 + bD_x^{-1} \circ c$$

is skew-adjoint. Then the operator δ in (8.4) acquires the form

$$\tilde{\delta}(\rho) = \frac{1}{2}w\left(\frac{\partial\rho}{\partial w} + \sum_{i\geq 0}(-1)^{i+1}D_x\left(\frac{1}{c}D_x^i\frac{\partial\rho}{\partial q_i}\right)\right) \tag{8.9}$$

for any quadratic in q_i and w function ρ. Thus condition (8.4) reads now

$$\tilde{\delta}(\mathbf{X}(q\psi)) = 0.$$

In the case of the pmKdV equation, ψ is given by (8.8), and we have

$$q\psi = q\left(q_{xxx} + \frac{2}{3}u_x^2 q_x + \frac{2}{3}u_x u_{xx} q - \frac{2}{3}u_{xx}w\right) = q\left(q_{xxx} + \frac{2}{3}u_x^2 q_x - \frac{2}{3}u_{xx}w\right)$$

and

$$\omega = \mathbf{X}(q\psi) = \frac{2}{3}q\left(2u_x q_x q_x - q_{xx}w - u_{xx}\,\mathbf{X}w\right) = \frac{2}{3}qq_{xx}w.$$

Thus, modulo a constant multiplier, one has

$$\begin{aligned}
\delta(\omega) \sim w\left(\frac{\partial\omega}{\partial w} - D_x\frac{1}{u_{xx}}\frac{\partial\omega}{\partial q} - D_x\frac{1}{u_{xx}}D_x^2\frac{\partial\omega}{\partial q_{xx}}\right) \\
= w\left(qq_{xx} - D_x\frac{1}{u_{xx}}q_{xx}w + D_x\frac{1}{u_{xx}}D_x^2 qw\right) \\
= w\left(qq_{xx} - D_x\frac{1}{u_{xx}}q_{xx}w + D_x\frac{1}{u_{xx}}(q_{xx}w + u_{xx}q_x q)\right) = w(qq_{xx} + q_{xx}q) = 0.
\end{aligned}$$

Consequently, $\mathscr{S}$ is a symplectic structure.

8.2 Examples

We begin computer-based examples with a non-evolution equation that admits a local symplectic structure.

8.2.1 The Two-Dimensional WDVV Equation

It is readily checked that for the simplest Witten-Dijkgraaf-Verlinde-Verlinde (WDVV) equation

$$u_{yyy} - u_{xxy}^2 + u_{xxx}u_{xyy} = 0,$$

the operator D_x is a symplectic structure. The tangent covering of this equation is defined by the equation

$$q_{yyy} - 2u_{xxy}q_{xxy} + u_{xxx}q_{xyy} + u_{xyy}q_{xxx} = 0.$$

The CDE implementation goes as follows. We can compute the linearization of the above equation and its adjoint as we did in Sect. 2.2.3. To this aim we load the program wdvv_ell1.red:

```
indep_var:={x,y};
dep_var:={u};
odd_var:={q};
total_order:=10;
cde({indep_var,dep_var,odd_var,total_order},{});
```

and the function that defines the WDVV equation:

```
f_wdvv:={u_3y - u_2xy**2 + u_3x*u_x2y};
```

Then

```
ell_function(f_wdvv,lwdvv);
adjoint_cdiffop(lwdvv,lwdvv_star);
```

where the result is

```
lwdvv_sf(1);

- 2*q_2xy*u_2xy + q_3x*u_x2y + q_3y + q_x2y*u_3x;

lwdvv_star_sf(1);

- q_2x*u_2x2y + 2*q_2xy*u_2xy - q_2y*u_4x
- q_3x*u_x2y - q_3y - q_x2y*u_3x + 2*q_xy*u_3xy;
```

Then, in the program file wdvv_sympl1.red, we load the equation and its tangent covering:

```
principal_der:={u_3y};
de:={u_2xy**2 - u_3x*u_x2y};
principal_odd:={q_3y};
de_odd:={2*u_2xy*q_2xy - q_3x*u_x2y - u_3x*q_x2y};
```

and call CDE Then we load the adjoint linearization

```
mk_superfun(lwdvv_star_sf,1,1);
lwdvv_star_sf(1):= - u_2x2y*q_2x + 2*u_2xy*q_2xy
 - u_3x*q_x2y + 2*u_3xy*q_xy - u_4x*q_2y
 - u_x2y*q_3x - q_3y;
conv_superfun2cdiff(lwdvv_star_sf,lwdvv_star);
```

and check that the operator q_x is a shadow of cosymmetry of a cotangent covering, i.e., that the following command returns zero:

```
lwdvv_star(1,1,q_x);
```

Of course, we might search for other local symplectic operators using CRACK in a similar way as we did for Hamiltonian operators for the KdV equation, but to our knowledge there are none, [62].

8.2.2 The Krichever-Novikov Equation

The Krichever-Novikov equation is

$$u_t = u_{xxx} - \frac{3}{2}\frac{u_{xx}^2}{u_x},$$

while

$$q_t = q_{xxx} - 3\frac{u_{xx}}{u_x}q_{xx} + \frac{3}{2}\frac{u_{xx}^2}{u_x^2}q_x,$$

defines the tangent covering. Solving the equation $\tilde{\ell}_{\mathscr{E}}^*(\psi) = 0$ on $\mathscr{T}\mathscr{E}$, one obtains the solution (the simplest one)

$$\psi = \frac{1}{u_x^2}q_1 - \frac{u_{xx}}{u_x^3}q_0$$

with the corresponding operator

$$\mathscr{S} = \frac{1}{u_x^2}D_x - \frac{u_{xx}}{u_x^3}.$$

Obviously, $\mathscr{S}$ is skew-adjoint and

$$\omega_{\mathscr{S}} = q_0\left(\frac{1}{u_x^2}q_1 - \frac{u_{xx}}{u_x^3}q_0\right) = \frac{1}{u_x^2}q_0q_1.$$

Consequently,

$$\mathbf{X}(\omega_{\mathscr{S}}) = -2\frac{1}{u_x^3}q_1q_0q_1 = 0,$$

and thus $\mathscr{S}$ is a variational symplectic structure.

The CDE approach goes as follows. After computing the linearization and its adjoint (program file kn_ell1.red), we load the equation and its tangent covering (program file kn_sympl1.red):

```
principal_der:={u_t};
de:={u_3x - (3/2)*((u_2x**2)/u_x)};
principal_odd:={q_t};
de_odd:={q_3x + (3/2)*((u_2x**2)/(u_x**2))*q_x
  - 3*(u_2x/u_x)*q_2x};
```

Then we load the adjoint linearization and convert it to the corresponding $\mathscr{C}$-differential operator:

```
mk_superfun(lkn_star_sf,1,1);
lkn_star_sf(1):=(6*q*u_2x**3 - 12*q*u_2x*u_3x*u_x
 + 6*q*u_4x*u_x**2 + 6*q_2x*u_2x*u_x**2
 + 2*q_3x*u_x**3 - 2*q_t*u_x**3 - 9*q_x*u_2x**2*u_x
 + 12*q_x*u_3x*u_x**2)/(2*u_x**3);
conv_superfun2cdiff(lkn_star_sf,lkn_star);
```

We define the candidate symplectic operator

```
mk_superfun(symp,1,1);
symp(1):=(1/(u_x**2))*q_x - (u_2x/(u_x**3))*q;
```

(it could have been found by a CRACK search). It is a shadow of symmetry on the cotangent covering, the following command yields 0:

```
lkn_star(1,1,symp(1));
0;
```

Then we compute the symplecticity condition. First of all we convert the generating function of the operator to a variational *two*-form. A variational k-form is represented in CDE by a one-component superfunction of k-th odd degree, vform_symp in our case:

```
conv_genfun2vform(symp,vform_symp);
```

In order to see the superfunction, do

```
vform_symp(1);

 ext(q,q_x)
------------
        2
     u_x
```

We compute the Cartan differential:

```
cartan_df(vform_symp,dc_vform_symp);
```

It is zero, as the output of the following command shows:

```
dc_vform_symp(1);
0;
```

We can directly compute the symplecticity test or the variational differential of the variational form, by

```
variational_df(vform_symp,vform_res);
```

The result is a variational *three*-form, or a one-component superfunction of 3-rd odd degree, which is zero in our case, as the following command shows:

```
vform_res(1);
0;
```

8.2.3 Korteweg-de Vries Equation

If we solve the equation

$$\ell_{\mathscr{E}}(\psi) \equiv D_x^3(\psi) + uD_x(\psi) - D_t(\psi) = 0$$

on $\mathscr{T}\mathscr{E}$ for $\psi = a_0 q + \cdots + a_k q_k$, the sole solution will be $\psi = 0$. Let us extend, like in Sect. 7, the tangent space by the odd nonlocal variable w defined by $w_x = q$. Then we have

$$\mathbf{X}(w_x) = D_x(\mathbf{X}(w)) = \mathbf{X}(q) = 0.$$

It is easy to show that the operator D_x has no kernel in the space of functions on $\mathscr{T}\mathscr{E}$ linear with respect to $w, q, \ldots, q_k, \ldots$ Thus,

$$\mathbf{X}w = 0.$$

In the extended setting, we obtain a nontrivial solution $\psi = w$ to which the operator

$$\mathscr{S} = D_x^{-1}$$

corresponds. Then $\omega_{\mathscr{S}} = qw$ and $\mathbf{X}(\omega_{\mathscr{S}}) = 0$. So, no additional check is needed, and D_x^{-1} is a nonlocal variational symplectic structure for the KdV equation.

8.2.4 Dispersionless Boussinesq System

After loading the tangent covering with the three nonlocal variables (Sect. 6.2.2), we run the program bou_sympl1.red whose initial part is the same as that of bou_ro1.red, that was described in Sect. 7.2.2. This time, we will change the equation, as we are looking for shadows of cosymmetries. We define a space of weighted monomials (with degrees ranging from 1 to 8) which are linear in the odd variables as follows:

```
l_grad_mon:=der_deg_ordering(0,all_parametric_der);
l_grad_odd:=
  {1} . der_deg_ordering(1,all_parametric_odd);
gradmon:=graded_mon(1,10,l_grad_mon);
gradmon:={1} . gradmon;
linodd:=mkalllinodd(gradmon,l_grad_odd,1,8);
```

An ansatz for the (nonlocal) symplectic operator is

```
operator phi;
for i:=1:nc do phi(i):=(for each el in linodd sum
  (c(ctel:=ctel+1)*el));
```

Then, we load the adjoint linearization, computed in Sect. 2.3.2, into the operator `lbou_star` and calculate the equations:

```
for i:=1:nc do
  equ(i):=for j:=1:nc sum lbou_star(i,j,phi(j));
```

Solving the equations in the same way as Sect. 7.2.2, we will get as a solution the nonlocal operator

```
phi(1) := c(4)*o2;
phi(2) := c(4)*o1;
phi(3) := c(4)*(2*o1*w + o3);
```

8.2.5 *Camassa-Holm Equation*

After loading the tangent covering (with the nonlocal odd variables) and the weights as in Sect. 6.2.3 (see the program file ch_ro1.red), we load the adjoint linearization (note that we are now in the program file ch_sympl1.red)

```
mk_superfun(lch_star_sf,1,1);
lch_star_sf(1):= - alpha*q_t - 3*alpha*q_x*u
 + q_2x*u_x + q_3x*u + q_t2x + q_x*u_2x;
conv_superfun2cdiff(lch_star_sf,lch_star);
```

and construct an ansatz of weights between 1 and 6:

```
l_grad_mon:=der_deg_ordering(0,all_parametric_der);
l_grad_odd:=der_deg_ordering(1,all_parametric_odd);
gradmon:=graded_mon(1,10,l_grad_mon);
gradmon:=part(gradmon,2):=alpha . part(gradmon,2);
gradmon:={1} . gradmon;
```

```
l_grad_odd:={1} . l_grad_odd;
linodd:=mkalllinodd(gradmon,l_grad_odd,1,6);
phi:=for each el in linodd sum c(ctel:=ctel+1)*el;
```

The equation for shadows of cosymmetries on the tangent covering is computed by

```
equ 1:=num lch_star(1,1,phi);
```

Solving the above algebraic equation yields

```
phi:=c(17)*alpha*o1 + c(8)*o1*u + c(8)*o2 - c(8)*q_t
  - c(8)*q_x*u + c(3)*o1;
```

This amounts to two nonlocal symplectic operators (one of them is repeated two times).

8.2.6 Multi-dimensional Examples

Let us pass to multi-dimensional systems.

8.2.6.1 The General Heavenly Equation

In the paper [25], the general heavenly equation was introduced:

$$\alpha u_{tx}u_{yz} + \beta u_{ty}u_{xz} + \gamma u_{tz}u_{xy} = 0, \qquad \alpha + \beta + \gamma = 0, \tag{8.10}$$

together with a Lax pair. In [137] several new recursion, symplectic and Hamiltonian operators have been found, starting from the remark that the general heavenly equation admits partner symmetries.

Here, we start from the evolutionary presentation of the above equation:

$$u_t = v, \quad v_t = \frac{1}{u_{yz}}\big(u_{xx}u_{yz} - u_{xy}u_{xz} + v_y v_z + b(v_y u_{xz} - v_z u_{xy})\big) \tag{8.11}$$

(where $b = (\beta - \gamma)/\alpha$), and we find a local symplectic operator for (8.11), namely,

$$K = \begin{pmatrix} b(u_{xz}D_y - u_{xy}D_z) + v_zD_y + v_yD_z + v_{yz} & -u_{yz} \\ u_{yz} & 0 \end{pmatrix} \tag{8.12}$$

This coincides with Equation (4.3) in [137, p. 128].

We describe the program file gh_sympl1.red. After loading the jet space

```
indep_var:={t,x,y,z};
dep_var:={u,v};
odd_var:={p,q};
total_order:=5;
```

we load the equation

```
principal_der:={u_t,v_t};
de:={v,
  (1/u_yz)*(u_2x*u_yz - u_xy*u_xz + v_y*v_z
  + b*(v_y*u_xz - v_z*u_xy))};
```

and its linearization as a vector function of odd variables (computed using the program file gh_ell1.red):

```
tc_gh:={p_t - q,
(b*p_xy*u_yz*v_z - b*p_xz*u_yz*v_y -
b*p_yz*u_xy*v_z + b*p_yz*u_xz*v_y - b*q_y*u_xz*u_yz
+ b*q_z*u_xy*u_yz - p_2x*u_yz**2 + p_xy*u_xz*u_yz
+ p_xz*u_xy*u_yz - p_yz*u_xy*u_xz + p_yz*v_y*v_z +
 q_t*u_yz**2 - q_y*u_yz*v_z - q_z*u_yz*v_y)/u_yz**2};
```

Solving by the t-derivatives of the odd variables yields the tangent covering:

```
tan_covering_eq:=first
  solve({part(tc_gh,1),part(tc_gh,2)},{p_t,q_t});
principal_odd:={p_t,q_t};
de_odd:=
  {rhs(first tan_covering_eq),
   rhs(second tan_covering_eq)};
```

Now, we make a polynomial ansatz for the symplectic operator: we require the operator (as a vector superfunction) to be separately linear with respect to both odd and even variables. We also require the order of derivatives to be not higher than 2. The constant b has to be taken into account either. To this aim, after initializing the counter of unknown constants as

```
ctel:=0;
operator c,equ;
nc:=length(dep_var);
```

we define the list of variables that we want in the ansatz:

```
even_vars:=for i:=0:2 join
  selectvars(0,i,dep_var,all_parametric_der);
all_vars:=1 . even_vars;
odd_vars:=for i:=0:1 join
  selectvars(1,i,odd_var,all_parametric_odd);
ansatz_mon:= for each el in odd_vars join
  for each ell in even_vars join {ell*el,b*ell*el};
```

The ansatz for the symplectic operator is constructed as

```
operator phi;
phi(1):=(for each el in ansatz_mon sum
  (c(ctel:=ctel+1)*el));
```

```
phi(2):=(for each el in ansatz_mon sum
  (c(ctel:=ctel+1)*el));
```

Then, we need the adjoint linearization operator, computed in gh_ell1.red. We load it as a superfunction:

```
lgh_star_sf(1):=(2*b*q*u_2y2z*u_xy*u_yz*v_z - 2*b*
q*u_2y2z*u_xz*u_yz*v_y - 2*b*q*u_2yz*u_x2z*u_yz*
v_y - 6*b*q*u_2yz*u_xy*u_y2z*v_z + 2*b*q*u_2yz*
u_xy*u_yz*v_2z + 4*b*q*u_2yz*u_xyz*u_yz*v_z + 6*b*
q*u_2yz*u_xz*u_y2z*v_y - 2*b*q*u_2yz*u_xz*u_yz*
v_yz - b*q*u_2yz*u_yz**2*v_xz + 2*b*q*u_x2y*u_y2z*
u_yz*v_z - b*q*u_x2y*u_yz**2*v_2z - 2*b*q*u_x2yz*
u_yz**2*v_z + b*q*u_x2z*u_yz**2*v_2y + 2*b*q*u_xy*
u_y2z*u_yz*v_yz - b*q*u_xy*u_yz**2*v_y2z + 2*b*q*
u_xy2z*u_yz**2*v_y - 4*b*q*u_xyz*u_y2z*u_yz*v_y -
2*b*q*u_xz*u_y2z*u_yz*v_2y + b*q*u_xz*u_yz**2*
v_2yz + b*q*u_y2z*u_yz**2*v_xy - b*q_x*u_2yz*u_yz
**2*v_z + b*q_x*u_y2z*u_yz**2*v_y + b*q_xy*u_yz**3
*v_z - b*q_xz*u_yz**3*v_y + b*q_y*u_x2z*u_yz**2*
v_y + 2*b*q_y*u_xy*u_y2z*u_yz*v_z - b*q_y*u_xy*
u_yz**2*v_2z - 2*b*q_y*u_xyz*u_yz**2*v_z - 2*b*q_y
*u_xz*u_y2z*u_yz*v_y + b*q_y*u_xz*u_yz**2*v_yz + b
*q_y*u_yz**3*v_xz - b*q_yz*u_xy*u_yz**2*v_z + b*
q_yz*u_xz*u_yz**2*v_y + 2*b*q_z*u_2yz*u_xy*u_yz*
v_z - 2*b*q_z*u_2yz*u_xz*u_yz*v_y - b*q_z*u_x2y*
u_yz**2*v_z - b*q_z*u_xy*u_yz**2*v_yz + 2*b*q_z*
u_xyz*u_yz**2*v_y + b*q_z*u_xz*u_yz**2*v_2y - b*
q_z*u_yz**3*v_xy - p_t*u_yz**4 - q*u_2xy*u_y2z*
u_yz**2 + 2*q*u_2xyz*u_yz**3 - q*u_2xz*u_2yz*u_yz
**2 + 2*q*u_2y2z*u_xy*u_xz*u_yz - 2*q*u_2y2z*u_yz*
v_y*v_z + 2*q*u_2yz*u_x2z*u_xy*u_yz - 6*q*u_2yz*
u_xy*u_xz*u_y2z + 4*q*u_2yz*u_xyz*u_xz*u_yz + 6*q*
u_2yz*u_y2z*v_y*v_z - 2*q*u_2yz*u_yz*v_2z*v_y - 2*
q*u_2yz*u_yz*v_yz*v_z - q*u_x2y*u_x2z*u_yz**2 + 2*
q*u_x2y*u_xz*u_y2z*u_yz - 2*q*u_x2yz*u_xz*u_yz**2
- 2*q*u_xy*u_xy2z*u_yz**2 + 4*q*u_xy*u_xyz*u_y2z*
u_yz - 3*q*u_xyz**2*u_yz**2 - 2*q*u_y2z*u_yz*v_2y*
v_z - 2*q*u_y2z*u_yz*v_y*v_yz + q*u_yz**2*v_2y*
v_2z + q*u_yz**2*v_2yz*v_z + q*u_yz**2*v_y*v_y2z +
 q*u_yz**2*v_yz**2 - q_2x*u_yz**4 - q_x*u_2yz*u_xz
*u_yz**2 - q_x*u_xy*u_y2z*u_yz**2 + 2*q_x*u_xyz*
u_yz**3 + q_xy*u_xz*u_yz**3 + q_xz*u_xy*u_yz**3 +
q_y*u_2xz*u_yz**3 - q_y*u_x2z*u_xy*u_yz**2 + 2*q_y
*u_xy*u_xz*u_y2z*u_yz - 2*q_y*u_xyz*u_xz*u_yz**2 -
 2*q_y*u_y2z*u_yz*v_y*v_z + q_y*u_yz**2*v_2z*v_y +
```

```
 q_y*u_yz**2*v_yz*v_z - q_yz*u_xy*u_xz*u_yz**2 +
q_yz*u_yz**2*v_y*v_z + q_z*u_2xy*u_yz**3 + 2*q_z*
u_2yz*u_xy*u_xz*u_yz - 2*q_z*u_2yz*u_yz*v_y*v_z -
q_z*u_x2y*u_xz*u_yz**2 - 2*q_z*u_xy*u_xyz*u_yz**2
+ q_z*u_yz**2*v_2y*v_z + q_z*u_yz**2*v_y*v_yz)/
u_yz**4;

lgh_star_sf(2):=( - b*q*u_2yz*u_xz + b*q*u_xy*
u_y2z + b*q_y*u_xz*u_yz - b*q_z*u_xy*u_yz - p*u_yz
**2 - q*u_2yz*v_z - q*u_y2z*v_y + 2*q*u_yz*v_yz -
q_t*u_yz**2 + q_y*u_yz*v_z + q_z*u_yz*v_y)/u_yz**2;
```

then we convert it into a $\mathscr{C}$-differential operator:

```
conv_superfun2cdiff(lgh_star_sf,lgh_star);
```

The equations for shadows of cosymmetries on the tangent covering on our ansatz is

```
for i:=1:nc do equ(i):=
  num(for j:=1:nc sum lgh_star(i,j,phi(j)));
```

After solving, we get

```
  phi(1)  := c(36)*b**2*p_y*u_xz
  - c(36)*b**2*p_z*u_xy + c(36)*b*p*v_yz
  + c(36)*b* p_y*v_z + c(36)*b*p_z*v_y
  - c(36)*b*q*u_yz + c(35)*b*p_y*u_xz
  - c(35)*b*p_z*u_xy + c(35)*p*v_yz
  + c(35)*p_y*v_z + c(35)*p_z*v_y
  - c(35)*q*u_yz;

phi(2)  := p*u_yz*(c(36)*b + c(35));
```

As it is easy to see, there are two solutions which are identical up to a factor b. But we still have to test skew-adjointness and the symplectic property for our solution. Let us load it into a $\mathscr{C}$-differential operator:

```
mk_superfun(symp1_sf,1,2);
for i:=1:nc do
  symp1_sf(i):=df(phi(i),c(35));
conv_superfun2cdiff(symp1_sf,symp1);
```

a simple skew-adjointness test is

```
adjoint_cdiffop(symp1,symp1_star);
for i:=1:nc do if
  symp1_star_sf(i) + symp1_sf(i) neq 0 then
  write "Non self-adjoint operator";
```

The symplectic property is verified as follows. We can compute the Cartan differential with the commands

```
cartan_df(vform1,dc_vform1);
```

with result

```
dc_vform1(1);
ext(p,p_y,p_xz)*b + ext(p,p_y,q_z)
- ext(p,p_z,p_xy)*b - 2*ext(p,q,p_yz)
- ext(p,q_y,p_z);
```

then we should check if the above expression is a total divergence. It is simpler to use the following function, that performs also this second computation:

```
variational_df(vform1,vform_res);
```

whose result is

```
vform_res(1);
0;
```

8.2.6.2 The Plebanski Equation

The Plebanski Eq. (6.6) is Lagrangian; hence its linearization is self-adjoint: $\ell_F = \ell_F^*$. This means that tangent and cotangent coverings are the same space and that any symplectic operator is to be a recursion one also.

In the paper [105], the authors derive a symplectic operator by first passing to an evolutionary form of the Plebanski equation (see also Sect. 7.2.5). However, in the original coordinates, the recursion operator that is exhibited in Sect. 7.2.5 is also a symplectic operator, and its expression could not be simpler.

The above considerations apply to a wide range of Lagrangian systems. One example is the general heavenly Eq. (8.10), and other examples lie in the class of systems of PDEs which was introduced in [9] starting from the Ovsienko system [111]

$$u_{ty} = u_{xy}u_x - u_{xx}u_y, \quad p_{ty} = 2(u_{xx}p_y - u_{xy}p_x) + u_x p_{xy} - u_y p_{xx} - 2(u_{xx}u_y + 2u_{xy}u_x). \tag{8.13}$$

In [9] the authors obtained several nonlocal bivectors that are candidate to be Hamiltonian or, identically, symplectic operators, for the Ovsienko system (8.13).

Chapter 9
Cotangent Covering

Abstract The cotangent covering of an equation $\mathcal{E}$ is a natural object to define and compute Hamiltonian operators and recursion operators for cosymmetries of $\mathcal{E}$. It is a counterpart of the cotangent bundle in the world of differential equations. In this chapter we give the solution to Problem 1.19.

9.1 General Theory

Let $\mathcal{E} \subset J^\infty(n,m)$ be an equation given by the condition $F = 0$, where $F = (F^1,\dots,F^r) \in \mathcal{F}(n,m;r) = P$. Recall that the equation $\mathcal{E}$ must satisfy the regularity condition: for any $\mathcal{C}$-differential operator Δ, the equality $\Delta(F) = 0$ implies $\Delta|_{\mathcal{E}} = 0$.

Let $\mathcal{E}$, in addition, be a normal equation (see p. 7). Its *cotangent covering* denoted by $\mathcal{T}^*\mathcal{E} \subset J^\infty(n,m+r)$ is given by the system

$$F^j(x,\dots,u_\sigma,\dots) = 0, \quad \ell^*_{F^j}(x,\dots,u_\sigma,\dots,p_\sigma,\dots) = 0, \qquad j = 1,\dots,r,$$

where $p = (p^1,\dots,p^r)$ is a new auxiliary dependent variable. Similar to the case of the tangent bundle (Sect. 6.1), this variable is odd.

Example 9.1 The system

$$\begin{aligned} u_t &= uu_x + u_{xx}, \\ p_t &= up_x - p_{xx} \end{aligned}$$

is the cotangent covering of the Burgers equation.

Example 9.2 The linearization operator of the Monge-Ampére equation

$$u_{xx}u_{yy} - u_{xy}^2 = 0$$

J. Krasil'shchik et al., *The Symbolic Computation of Integrability Structures for Partial Differential Equations*, Texts and Monographs in Symbolic Computation,
https://doi.org/10.1007/978-3-319-71655-8_9

is of the form

$$\ell_{\mathcal{E}} = u_{xx}D_y^2 + u_{yy}D_x^2 - 2u_{xy}D_xD_y.$$

This is a self-adjoint $\mathcal{C}$-differential operator, and thus the cotangent covering

$$u_{xx}u_{yy} - u_{xy}^2 = 0,$$
$$u_{xx}p_{yy} + u_{yy}p_{xx} - 2u_{xy}p_{xy} = 0$$

coincides with the tangent one.

Example 9.3 The Drinfel′d-Sokolov system is of the form

$$\begin{aligned} u_t &= 3vv_x, \\ v_t &= 2v_{xxx} + u_xv + 2uv_x. \end{aligned} \tag{9.1}$$

Its linearization is the matrix $\mathcal{C}$-differential operator

$$\ell_{\mathcal{E}} = \begin{pmatrix} D_t & -3vD_x - 3v_x \\ -vD_x - 2v_x & D_t - 2D_x^3 - 2uD_x - u_x \end{pmatrix}.$$

Consequently, the adjoint one is of the form

$$\ell_{\mathcal{E}}^* = \begin{pmatrix} -D_t & vD_x - v_x \\ 3vD_x & -D_t + 2D_x^3 + 2uD_x + u_x \end{pmatrix},$$

and the cotangent covering is obtained from the system (9.1) by adding another two equations

$$\begin{aligned} p_t^u &= vp_x^v - v_xp^v, \\ p_t^v &= 3vp_x^u + 2p_{xxx}^v + 2up_x^v + u_xp^v. \end{aligned}$$

The following two properties of the cotangent covering are essential for us (cf. Sect. 6.1):

(1) there exists a one-to-one correspondence between the maps $\psi\colon \mathcal{E} \to \mathcal{T}^*\mathcal{E}$ whose differential takes the total derivatives on $\mathcal{E}$ to linear combinations of those on $\mathcal{T}^*\mathcal{E}$ and cosymmetries of $\mathcal{E}$;
(2) there is a canonical one-to-one correspondence between symmetries of $\mathcal{E}$ and conservation laws of $\mathcal{T}^*\mathcal{E}$ linear with respect to p_σ.

Remark 9.1 Similar to what was said in Remark 6.3 about $\mathcal{T}\mathcal{E}$, Property 2 is due to the Green formula, and in the case of evolution equations in t and x, the x-component of the conservation law $\omega(\varphi)$ corresponding to a symmetry $\varphi = (\varphi^1, \dots, \varphi^n)$ equals $\sum_{j=1}^m \varphi^j p^j$.

Conservation laws on $\mathcal{T}^*\mathcal{E}$ that correspond to symmetries of $\mathcal{E}$ are called *canonical*.

Example 9.4 Take the Burgers equation and its symmetry $\varphi = u_x$. Then, the corresponding canonical conservation law is

$$\omega_\varphi = u_x p\,dx + \big((uu_x + u_{xx})p - u_x p_x\big)\,dt.$$

Example 9.5 Let us transform the Monge-Ampére equation to the evolutionary form:

$$\begin{aligned} u_y &= v, \\ v_y &= \frac{v_x^2}{u_{xx}}. \end{aligned} \tag{9.2}$$

Then, the cotangent covering is obtained by adding the system

$$\begin{aligned} p_y^u &= \left(\frac{v_x^2}{u_{xx}^2} p^v\right)_{xx}, \\ p_y^v &= 2\left(\frac{v_x}{u_{xx}} p^v\right)_x + p^u. \end{aligned}$$

Equation (9.2) possesses the symmetry $\varphi = (1, 0)$, and the corresponding conservation law on $\mathcal{T}^*\mathcal{E}$ is

$$\omega_\varphi = p^u\,dx + \left(\frac{v_x^2}{u_{xx}^2} p^v\right)_x dy.$$

Example 9.6 The Drinfel′d-Sokolov system is x-invariant, i.e., admits the symmetry $\varphi = (u_x, v_x)$. The associated form

$$\omega_\varphi = (u_x p^u + v_x p^v)\,dx + \big(3v v_x p^u + (2v_{xxx} + u_x v + 2uv_x)p^v + v_x p_{xx}^v - v_{xx} p_V^v\big)\,dt$$

is a canonical conservation law on $\mathcal{T}^*\mathcal{E}$.

Note that for any normal equation $\mathcal{E}$, its cotangent covering $\mathcal{T}^*\mathcal{E}$ is always a Lagrangian equation: if $\mathcal{E}$ is given by a vector function $(F^1, \dots, F^r)$, the corresponding Lagrangian is $\mathcal{L} = p^1 F^1 + \dots + p^r F^r$.

Again, in parallel with Sect. 6.1, we shall also use the following construction. Let $\Delta\colon \mathcal{F}(\mathcal{E}; r) \to \mathcal{F}(\mathcal{E}; l)$ be a $\mathcal{C}$-differential operator given by the matrix $\left(\Delta_i^j\right)$

$$\Delta_i^j = \sum_\sigma b_{i,\sigma}^j D_\sigma, \qquad i = 1, \dots, r, \quad j = 1, \dots, l.$$

Then, we construct a vector function $\psi_\Delta = (\psi_\Delta^1, \ldots, \psi_\Delta^l)$ on $\mathcal{T}^*\mathcal{E}$ of the form

$$\psi_\Delta^j = \sum_{i,\sigma} b_{i,\sigma}^j p_\sigma^i;$$

the function is odd.

9.2 Examples

When one works with the tangent covering at the level of the CDE implementation, nothing changes with respect to the tangent covering: CDE regards both tangent and cotangent covering as a system of PDEs in even variables with a system of PDEs which is linear in odd variables "on top" of it. The syntax for generating both coverings is the same, and adding nonlocal variables works in both cases exactly in the same way.

9.2.1 Korteweg-de Vries Equation

The cotangent covering of the KdV equation is

$$u_t = uu_x + u_{xxx},$$
$$p_t = up_x + p_{xxx}.$$

The Lagrangian for $\mathcal{T}^*\mathcal{E}$ is $\mathcal{L} = p(uu_x + u_{xxx} - u_t)$. To the symmetry $\varphi = u_x$ of the KdV equation, there corresponds the conservation law

$$\omega(\varphi) = u_x p\, dx + \big((uu_x + u_{xxx})p - u_{xx}p_x + u_x p_{xx}\big)\, dt \tag{9.3}$$

of $\mathcal{T}^*\mathcal{E}$.

Let us consider, for example, the initial part of the program file kdv_ho4.red. Here, the cotangent covering of the KdV equation with the above nonlocal odd variable (9.3) is generated as follows:

```
indep_var:={x,t};
dep_var:={u};
odd_var:={p,r1};
total_order:=8;
principal_der:={u_t};
de:={u*u_x+u_3x};
principal_odd:={p_t,r1_x,r1_t};
de_odd:={u*p_x+p_3x,
```

```
p*u_x,
p*u*u_x + p*u_3x + p_2x*u_x - p_x*u_2x};
```

and consistent weights can be given as

```
deg_indep_var:={-1,-3};
deg_dep_var:={2};
deg_odd_var:={2,4};
```

9.2.2 Dispersionless Boussinesq System

In this case, the cotangent covering is

$$
\begin{aligned}
w_t &= u_x, & p_t^w &= wp_x^u - p_x^v + 2u_x p^v, \\
u_t &= ww_x + v_x, & p_t^u &= p_x^w - 3wp_x^v - 2w_x p^v, \\
v_t &= -uw_x - 3wu_x, & p_t^v &= p_x^u.
\end{aligned}
$$

The conservation law $\omega(\varphi) = X\,dx + T\,dt$ of $\mathcal{T}^*\mathcal{E}$ corresponding to a symmetry $\varphi = (\varphi^w, \varphi^u, \varphi^v)$ is of the form

$$
X = p^w\varphi^w + p^u\varphi^u + p^v\varphi^v, \quad T = p^w\varphi^u + p^u(\varphi^v + \varphi^w w) - p^v(3\varphi^u w + \varphi^w u).
$$

For example, the conservation law

$$
\omega = (p^w w_x + p^u u_x + p^v v_x)\,dx + \big(p^w u_x + p^u(v_x + w_x w) - p^v(3u_x w + w_x u)\big)\,dt
$$

corresponds to the x-translation $\varphi = (w_x, u_x, v_x)$.

In CDE the cotangent covering (without additional odd variables) is generated in the program file bou_ho1.red as follows:

```
indep_var:={x,t};
dep_var:={u,v,w};
odd_var:={p,q,r};
total_order:=8;
principal_der:={u_t,v_t,w_t};
de:={w*w_x + v_x, - u*w_x - 3w*u_x,u_x};
principal_odd:={p_t,q_t,r_t};
de_odd:={
 - 2*q*w_x - 3*q_x*w + r_x,
p_x,
p_x*w + 2*q*u_x - q_x*u
  };
```

Nonlocal variables can be added to both the even and the odd system of equations; see [57, 61] for examples of nonlocal Hamiltonian operator for the dispersionless Boussinesq equation (see the next chapter for the discussion).

9.2.3 Camassa-Holm Equation

The cotangent covering is given by

$$\begin{aligned} &\alpha u_t - u_{txx} + 3\alpha u u_x = 2u_x u_{xx} + u u_{xxx}, \\ &-\alpha p_t + p_{xxt} + u p_{xxx} + u_x pxx + (u_{xx} - 3\alpha u)p_x = 0. \end{aligned}$$

The conservation law $\omega_\varphi = X\,dx + T\,dt$ corresponding to any symmetry φ of the Camassa-Holm equation has the form

$$\begin{aligned} X &= (\alpha\varphi - D_x^2(\varphi))p, \\ T &= ((u_{xx} - 3\alpha u)\varphi + u_x D_x(\varphi) + u D_x^2(\varphi))p - u D_x(\varphi)p_x + u\varphi p_{xx} - D_x(\varphi)p_t + \varphi p_{xt}. \end{aligned}$$

For example, the conservation law corresponding to the x-translation $\varphi = u_x$ has the components $X = (\alpha u_x - u_{xxx})p$ and

$$T = \big((2u_{xx} - 3\alpha u)u_x + u u_{xxx}\big)p - u u_{xx} p_x + u u_x p_{xx} - u_{xx} p_t + u_x p_{xt}.$$

In CDE, load the cotangent covering with the above nonlocal odd variable as follows (program file ch_ho1.red):

```
indep_var:={t,x};
dep_var:={u};
odd_var:={p,o1};
total_order:=8;
principal_der:={u_3x};
de:={(alpha*(u_t + 3*u*u_x) - u_t2x - 2*u_x*u_2x)/u};

phi:=u_x;
% The nonlocal odd variable
o1x:=(u_x*alpha - first(de))*p;
o1t:=
   ((u_2x - 3*u*alpha)*u_x + u_x*u_2x + u*first(de))*p
   - u*u_2x*p_x + u*u_x*p_2x - u_2x*p_t + u_x*p_tx;

principal_odd:={p_3x,o1_x,o1_t};
de_odd:={(alpha*p_t + 3*alpha*p_x*u - p_2x*u_x
  - p_t2x - p_x*u_2x)/u,o1x,o1t};
```

We can assign the weights

```
deg_indep_var:={-2,-1};
deg_dep_var:={1};
deg_odd_var:={1,4};
```

9.2.4 Multi-dimensional Examples

Consider two multi-dimensional examples briefly.

9.2.4.1 The Kadomtsev-Petviashvili Equation

The cotangent covering of the Kadomtsev-Petviashvili equation can be loaded for both the evolutionary and the non-evolutionary form. Let us start by the evolutionary form (2.14) (program file kp_ho1.red); we have

```
indep_var:={t,x,y};
dep_var:={u};
odd_var:={p};
total_order:=6;
principal_der:={u_2y};
de:={u_tx-u_x**2-u*u_2x-(1/12)*u_4x};
principal_odd:={p_2y};
de_odd:={p_tx-u*p_2x-(1/12)*p_4x};
```

Weights can be assigned as

```
deg_indep_var:={-3,-1,-2};
deg_dep_var:={2};
deg_odd_var:={1};
```

In the evolutionary case (2.15) (program file kpev_ho1.red), we have

```
indep_var:={t,x,y};
dep_var:={u,v};
odd_var:={p,q};
total_order:=6;
principal_der:={u_y,v_y};
de:={v_x,u_t - u*u_x - (1/12)*u_3x};
principal_odd:={p_y,q_y};
de_odd:={( - q_3x + 12*q_t - 12*q_x*u)/12,p_x};
```

Here, weights can be assigned by

```
deg_indep_var:={-3,-1,-2};
deg_dep_var:={2,3};
deg_odd_var:={2,1};
```

9.2.4.2 The Plebanski Equation

We already remarked that the tangent and the cotangent coverings coincide as the linearization of the equation is self-adjoint: $\ell_{\mathcal{E}} = \ell_{\mathcal{E}}^*$, or, equivalently, the equation is Lagrangian. (See Sects. 2.3.4.2 and 6.2.4.2).

Chapter 10
Variational Poisson Structures

Abstract A variational Poisson structure on a differential equation $\mathcal{E}$ is a $\mathcal{C}$-differential operator that takes cosymmetries of $\mathcal{E}$ to its symmetries and possesses the necessary integrability properties. In the literature on integrable systems, Poisson structures are traditionally called Hamiltonian operators. We expose here the computational theory of local variational Poisson structures for normal equations. In this chapter the solutions of Problems 1.24, 1.25, 1.26, and 1.28 is presented.

10.1 General Theory

Let $\mathscr{P}: \mathscr{F}(\mathcal{E}; r) = \hat{P} \to \mathscr{F}(\mathcal{E}; m) = \varkappa$, where $\mathcal{E} = \{F = 0\}$, $F \in P$, be a $\mathcal{C}$-differential operator of the form $\mathscr{P} = (\mathscr{P}_i^j)$ with

$$\mathscr{P}_i^j = \sum_\sigma b_{\sigma,i}^j D_\sigma, \qquad i = 1, \dots, r, \quad j = 1, \dots, m.$$

Using the construction of Sect. 9.1, let us consider the vector function $\varphi_{\mathscr{P}} = (\varphi_{\mathscr{P}}^1, \dots, \varphi_{\mathscr{P}}^m)$ on $\mathcal{T}^*\mathcal{E}$, where

$$\varphi_{\mathscr{P}}^j = \sum_{i,\sigma} b_{\sigma,i}^j p_\sigma^i. \tag{10.1}$$

Here again $\varphi_{\mathscr{P}}^i$ is an odd function.

Let $\tilde{\ell}_{\mathcal{E}}$ be the lift of the operator $\ell_{\mathcal{E}}$ to $\mathcal{T}^*\mathcal{E}$. Consider the equation:

$$\tilde{\ell}_{\mathcal{E}}(\varphi_{\mathscr{P}}) = 0. \tag{10.2}$$

Since the cotangent covering is a Δ-covering with $\Delta = \ell_{\mathcal{E}}^*$, solutions of (10.2) are in one-to-one correspondence with classes of $\mathcal{C}$-differential operators

$$\frac{\{\, \mathscr{P} \mid \ell_{\mathcal{E}} \circ \mathscr{P} = \Delta \circ \ell_{\mathcal{E}}^* \,\}}{\{\, \mathscr{P} \mid \mathscr{P} = \nabla \circ \ell_{\mathcal{E}}^* \,\}},$$

J. Krasil'shchik et al., *The Symbolic Computation of Integrability Structures for Partial Differential Equations*, Texts and Monographs in Symbolic Computation,
https://doi.org/10.1007/978-3-319-71655-8_10

where $\Delta\colon \mathscr{F}(\mathscr{E};m) \to \mathscr{F}(\mathscr{E};r)$ and $\nabla\colon \mathscr{F}(\mathscr{E};m) \to \mathscr{F}(\mathscr{E};m)$ are some $\mathscr{C}$-differential operators (see Proposition 1.2). Any operator $\mathscr{P}$ of such a kind takes cosymmetries of $\mathscr{E}$ to its symmetries.

Having an operator of this type, one can define a bracket on the set of conservation laws of $\mathscr{E}$:

$$\{\omega, \omega'\}_{\mathscr{P}} = L_{\psi_\omega}(\omega'), \tag{10.3}$$

where ψ_ω is the generating function of ω and L stands for the Lie derivative. Then $\mathscr{P}$ is a *variational Poisson structure* (or a *Hamiltonian operator*) for the equation $\mathscr{E}$ if this bracket

(1) is skew-symmetric and
(2) enjoys the Jacobi identity.

Then (10.3) is the *Poisson bracket* associated with the Poisson structure $\mathscr{P}$.

Example 10.1 Let $\mathscr{E}$ be a Lagrangian equation. Then the operator $\ell_{\mathscr{E}}$ is self-adjoint, and consequently the identical operator is a Poisson (and symplectic) structure on $\mathscr{E}$. In particular, since $\mathscr{T}^*\mathscr{E}$ is a Lagrangian equation for any $\mathscr{E}$, it always possesses a canonical Poisson and symplectic structure.

Properties 1–2 are guaranteed, first, by self-adjointness of the composition $\ell_{\mathscr{E}} \circ \mathscr{P}$:

$$\ell_{\mathscr{E}} \circ \mathscr{P} = \mathscr{P}^* \circ \ell_{\mathscr{E}}^* \tag{10.4}$$

and by another feature of $\mathscr{P}$ for which we need the notion of the variational Schouten bracket.

Consider two $\mathscr{C}$-differential operators $\mathscr{P}_1$ and $\mathscr{P}_2$ that satisfy Eq. (10.4), and using the regularity condition from Sect. 2.1.1, note that

$$\ell_F \circ \bar{\mathscr{P}}_i - \bar{\mathscr{P}}_i^* \circ \ell_F^* = \nabla_i(F, \cdot), \qquad i = 1, 2,$$

where $\bar{\mathscr{P}}_i$ is an arbitrary extension of the operator $\mathscr{P}_i$ from $\mathscr{E}$ to the ambient jet space $J^\infty(n,m)$ and $\nabla_i\colon \mathscr{F}(n,m;r) \times \mathscr{F}(n,m;r) \to \mathscr{F}(n,m;r)$ is $\mathscr{C}$-differential operator with respect to both arguments. Then their *variational Schouten bracket*

$$[\![\mathscr{P}_1, \mathscr{P}_2]\!]\colon \mathscr{F}(\mathscr{E};r) \times \mathscr{F}(\mathscr{E};r) \to \mathscr{F}(\mathscr{E};m)$$

is defined by

$$\begin{aligned}
[\![\mathscr{P}_1, \mathscr{P}_2]\!](\psi_1, \psi_2) = {} & \mathbf{E}_{\mathscr{P}_1(\psi_2)}(\mathscr{P}_2)(\psi_1) - \mathbf{E}_{\mathscr{P}_1(\psi_1)}(\mathscr{P}_2)(\psi_2) \\
& + \mathbf{E}_{\mathscr{P}_2(\psi_2)}(\mathscr{P}_1)(\psi_1) - \mathbf{E}_{\mathscr{P}_2(\psi_1)}(\mathscr{P}_1)(\psi_2) \\
& - \mathscr{P}_2(\nabla_1|_{\mathscr{E}}^{*_1}(\psi_1, \psi_2)) - \mathscr{P}_1(\nabla_2|_{\mathscr{E}}^{*_1}(\psi_1, \psi_2)),
\end{aligned} \tag{10.5}$$

where $\psi_1, \psi_2 \in \mathcal{F}(\mathcal{E}; r)$, and ^{*_1} denote, as in Sect. 8.1, conjugation with respect to the first argument. A $\mathcal{C}$-differential operator satisfying Eq. (10.4) is called a *variational Poisson structure* (or a *Hamiltonian operator*) if $[\![\mathcal{P}, \mathcal{P}]\!] = 0$. Two Poisson structures are *compatible* if their Schouten bracket vanishes.

Proposition 10.1 *If* $[\![\mathcal{P}, \mathcal{P}]\!] = 0$ *and Eq.* (10.4) *is fulfilled, then the bracket* $\{\cdot,\cdot\}_{\mathcal{P}}$ *given by* (10.3) *satisfies the Jacobi identity.*

Thus, to check whether an operator $\mathcal{P}$ that satisfies (10.4) is a Poisson structure on $\mathcal{E}$, we must prove that

$$\mathbf{E}_{\mathcal{P}(\psi_2)}(\mathcal{P})(\psi_1) - \mathbf{E}_{\mathcal{P}(\psi_1)}(\mathcal{P})(\psi_2) - \mathcal{P}(\nabla|_{\mathcal{E}}^{*_1}(\psi_1, \psi_2)) = 0 \tag{10.6}$$

is valid for any cosymmetries ψ_1 and ψ_2. In the case of evolution equations, the above condition simplifies drastically. Namely, if the function $\varphi_{\mathcal{P}}$ enjoys Eq. (10.2), then Eq. (10.4) amounts now to $\mathcal{P}^* = -\mathcal{P}$, while (10.6) may be rewritten as follows. Consider the vector function $\varphi_{\mathcal{P}}$ given by (10.1) and construct the quadratic function:

$$\Phi_{\mathcal{P}} = \sum_j \varphi^j_{\mathcal{P}} p^j = \sum_{\sigma,i,j} b^j_{\sigma,i} p^i_\sigma p^j.$$

Then condition (10.6) is equivalent to

$$\delta \sum_j \frac{\delta \Phi_{\mathcal{P}}}{\delta u^j} \frac{\delta \Phi_{\mathcal{P}}}{\delta p^j} = 0 \tag{10.7}$$

where δ is the Euler operator on $\mathcal{T}^*\mathcal{E}$ (see Sect. 4.1.1), while compatibility of two structures is expressed by the formula

$$\delta \sum_j \left(\frac{\delta \Phi_{\mathcal{P}_1}}{\delta u^j} \frac{\delta \Phi_{\mathcal{P}_2}}{\delta p^j} + \frac{\delta \Phi_{\mathcal{P}_2}}{\delta u^j} \frac{\delta \Phi_{\mathcal{P}_1}}{\delta p^j} \right) = 0. \tag{10.8}$$

Remark 10.1 The skew-adjointness of $\mathcal{P}$ can also be expressed in terms of $\Phi_{\mathcal{P}}$: it amounts to

$$\sum_j \frac{\delta \Phi_{\mathcal{P}}}{\delta p^j} = -2\Phi_{\mathcal{P}}. \tag{10.9}$$

Remark 10.2 As it follows from [60], if $\mathcal{E} = \{u_t = f(x, t, \dots, u_k, \dots)\}$ is an evolution equation of order >1 and the symbol of the right-hand side is nondegenerate, then condition (10.9) guarantees condition (10.7) as well, provided $\varphi_{\mathcal{P}}$ satisfies Eq. (10.2). Moreover, any two Poisson structures on such equations are compatible.

Example 10.2 Consider the Drinfel′d-Sokolov system

$$u_t = 3vv_x, \quad v_t = 2v_{xxx} + u_x v + 2uv_x$$

from Example 9.3 and the equation

$$\tilde{\ell}_{\mathcal{E}}(\varphi) = 0 \tag{10.10}$$

on $\mathcal{T}^*\mathcal{E}$. Then it is not difficult to find out that the vector function

$$\varphi = \begin{pmatrix} 2p_3^u + 2up_1^u + u_1p^u + 2vp_1^v + v_1p^v \\ 2vp_1^u + v_1p^u + 2p_3^v + 2up_1^v + u_1p^v \end{pmatrix}$$

is a solution of (10.10). Following the general scheme, construct the function

$$\begin{aligned} \Phi &= p^u\varphi^u + p^v\varphi^v \\ &= p^u(2p_3^u + 2up_1^u + u_1p^u + 2vp_1^v + v_1p^v) + p^v(2vp_1^u + v_1p^u + 2p_3^v + 2up_1^v + u_1p^v) \\ &= 2\big(p^u(p_3^u + up_1^u + vp_1^v) + p^v(vp_1^u + p_3^v + up_1^v)\big), \end{aligned}$$

where the canceled terms are due to the oddness of p_i^u and p_i^v. The operator corresponding to φ,

$$\mathcal{P}_\varphi = \begin{pmatrix} 2D_x^3 + 2uD_x + u_1 & 2vD_x + v_1 \\ 2vD_x + v_1 & 2D_x^3 + 2uD_x + u_1 \end{pmatrix}, \tag{10.11}$$

is skew-adjoint, but we cannot apply the result mentioned in Remark 10.2, because the symbol of the right-hand side of the Drinfel′d-Sokolov system is degenerate. Hence, we need a direct check of condition (10.7). Let $\Phi' = \Phi/2$. One obviously has

$$\frac{\delta\Phi'}{\delta u} = p^u p_1^u + p^v p_1^v, \qquad \frac{\delta\Phi'}{\delta v} = p^u p_1^v + p^v p_1^u.$$

Further, using again the oddness of the p-variables, we obtain

$$\frac{\delta\Phi'}{\delta p^u} = (p_3^u + up_1^u + vp_1^v) - D_x^3(p^u) - D_x(up^u) - D_v(vp^v) = -u_1p^u - v_1p^v$$

and

$$\frac{\delta\Phi'}{\delta p^v} = -D_x(vp^u) + (vp_1^u + up_1^v + p_3^v) - D_x(up^v) - D_x^3(p^v) = -v_1p^u - u_1p^v.$$

Consequently,

$$\begin{aligned} &\frac{\delta\Phi'}{\delta u}\cdot\frac{\delta\Phi'}{\delta p^u} + \frac{\delta\Phi'}{\delta v}\cdot\frac{\delta\Phi'}{\delta p^v} \\ &\quad = -(p^up_1^v + p^vp_1^u)(u_1p^u + v_1p^v) - (p^up_1^v + p^vp_1^u)(v_1p^u + u_1p^v) \\ &\quad = -(v_1p^up_1^up^v + u_1p^vp_1^vp^u + u_1p^up_1^vp^v + v_1p^vp_1^up^u) = 0, \end{aligned}$$

i.e., the operator (10.11) is a Poisson structure.

Example 10.3 The Harry Dym equation reads

$$u_t = u^3 u_{xxx}.$$

Its linearization is

$$\ell_{\mathscr{E}} = D_t - u^3 D_x^3 - 3uu^2 u_{xxx}$$

with the adjoint operator of the form

$$\begin{aligned} -D_t + D_x^3 \circ u^3 - 3uu^2 u_{xxx} \\ = -D_t + 6u_x^3 + 18uu_x u_{xx} + (18uu_x^2 + 9u^2 u_{xx})D_x + 9u^2 u_x D_x^2 + u^3 D_x^3. \end{aligned}$$

Consequently, the cotangent covering is defined by the additional equation

$$p_t = (6u_x^3 + 18uu_x u_{xx})p + (18uu_x^2 + 9u^2 u_{xx})p_x + 9u^2 u_x p_{xx} + u^3 p_{xxx}.$$

Solving the equation

$$\tilde{\ell}_{\mathscr{E}}(\varphi) = 0$$

on $\mathscr{T}^*\mathscr{E}$ with φ linear in the variables p_i, we get a solution of the form

$$\varphi = (6u^3 u_x^3 + 18u^4 u_x u_{xx})p + (18u^4 u_x^2 + 9u^2 u_{xx})p_x + 9u^5 u_x p_{xx} + u^6 p_{xxx}$$

to which the operator

$$\mathscr{P} = (6u^3 u_x^3 + 18u^4 u_x u_{xx}) + (18u^4 u_x^2 + 9u^2 u_{xx})D_x + 9u^5 u_x D_x^2 + u^6 D_x^3 = u^3 D_x^3 \circ u^3.$$

This operator is skew-adjoint, and since the symbol of right-hand is nondegenerate, it is a Poisson structure for the Harry Dym equation.

Consider nonlocal examples now.

Remark 10.3 We do not have well-defined techniques to check Condition 2 for nonlocal operators $\mathscr{P}$ at the moment. So, in the examples below we only *construct* operators of this type.

Example 10.4 Consider the Cavalcante-Tenenblat equation

$$u_t = \left(u_x^{-\frac{1}{2}}\right)_{xx} + u_x^{\frac{3}{2}}.$$

Its linearization is equivalent to the operator

$$u_x^{\frac{7}{2}} D_t - \left(\frac{3}{4}u_x u_{xxx} + \frac{3}{2}u_x^4 - \frac{15}{8}u_{xx}^2\right) D_x - \frac{3}{2}u_x u_{xx} D_x^2 + \frac{1}{2}u_x^2 D_x^3,$$

while the cotangent covering is defined by the equation

$$u_x^{\frac{5}{2}} p_t = \frac{3}{4} u_{xx} u_x^2 p + \frac{3}{2} u_x^3 p_x + \frac{3}{4} u_{xx} p_{xx} - \frac{1}{2} u_x p_{xxx}.$$

The equation $\tilde{\ell}_{\mathscr{E}}(\varphi) = 0$ has no nontrivial solution on $\mathscr{T}^*\mathscr{E}$, but if we add the canonical nonlocal variable w defined by

$$w_x = u_x w$$

and associated with the symmetry u_x, we shall find the solution

$$\varphi = p_x - u_x w,$$

to which the skew-adjoint operator

$$\mathscr{P} = D_x - u_x D_x^{-1} \circ u_x$$

corresponds.

Example 10.5 The fifth-order evolution equation

$$u_t = u_{xxxxx} + 5u_x u_{xxx} + \frac{5}{3} u_x^3$$

is called the potential Sawada-Kotera equation. Its linearization is

$$\ell_{\mathscr{E}} = D_t - 5(u_{xxx} + u_x^2) D_x - 5u_x D_x^3 - D_x^5$$

and the cotangent covering is defined by

$$p_t = 10(u_{xxxx} + u_x u_{xx}) p + 5(4u_{xxx} + u_x^2) p_x + 15 u_{xx} p_{xx} + 5 u_x p_{xxx} + p_{xxxxx}.$$

Let us introduce the canonical nonlocal variables on $\mathscr{T}^*\mathscr{E}$ associated with the symmetries

$$\varphi_1 = 1, \qquad \varphi_2 = u_x$$

and defined by the relations

$$w_x^1 = p, \qquad w_x^2 = u_x p.$$

The equation $\tilde{\ell}_{\mathscr{E}}(\varphi) = 0$ possesses a nontrivial solution in this extension. The solution is

$$\varphi = p_x + 2(u_x w^1 + w^2),$$

and the operator

$$\mathscr{P}_\varphi = D_x + 2(u_x D_x^{-1} + D_x^{-1} \circ u_x)$$

corresponds to this solution.

10.2 Examples

Let us now pass to the `CDE` implementation.

10.2.1 Korteweg-de Vries Equation

Let $\varphi = \sum_k b_k p_k$. Solving the equation

$$\ell_{\mathscr{E}}(\varphi) \equiv D_x^3(\varphi) + uD_x(\varphi) + u_x\varphi - D_t(\varphi) = 0 \tag{10.12}$$

on $\mathscr{T}^*\mathscr{E}$, we get two independent solutions

$$\varphi_1 = p_x, \qquad \varphi_2 = p_{xxx} + \frac{2}{3}up_x + \frac{1}{3}u_x p$$

to which the operators

$$\mathscr{P}_1 = D_x, \qquad \mathscr{P}_2 = D_x^3 + \frac{2}{3}uD_x + \frac{1}{3}u_x$$

correspond, both of them being skew-adjoint. The corresponding quadratic functions are

$$\Phi_1 = p_1 p, \qquad \Phi_2 = \Big(p_3 + \frac{2}{3}up_1\Big)p. \tag{10.13}$$

Since $\delta\Phi_1/\delta u = 0$, the operator $\mathscr{P}_1$ is a Poisson structure in an obvious way. For the second operator, one has

$$\frac{\delta\Phi_2}{\delta u} = \frac{2}{3}p_1 p, \qquad \frac{\delta\Phi_2}{\delta p} = -2\Big(p_3 + \frac{2}{3}up_1 + \frac{1}{3}u_1 p\Big) \tag{10.14}$$

and

$$\frac{\delta\Phi_2}{\delta u}\frac{\delta\Phi_2}{\delta p} = \frac{4}{3}pp_1p_3. \tag{10.15}$$

Consequently,

$$\delta\Big(\frac{\delta\Phi_2}{\delta u}\frac{\delta\Phi_2}{\delta p}\Big) = \frac{4}{3}\frac{\delta(pp_1p_3)}{\delta p} = \frac{4}{3}\big(p_1p_3 + D_x(pp_3) - D_x^3(pp_1)\big) = 0. \qquad (10.16)$$

Thus, $\mathscr{P}_2$ is a variational Poisson structure as well, and it is straightforward to show that $\mathscr{P}_1$ and $\mathscr{P}_2$ are compatible. One can also prove that all solutions of Eq. (10.12) are exhausted by $\mathbb{R}$-linear combinations of Φ_1 and Φ_2.

Nevertheless, if we extend $\mathscr{T}^*\mathscr{E}$ by the nonlocal variable w introduced in Sect. 9.2.1 and defined by $w_x = u_x p$, we shall obtain another solution:

$$\Phi_3 = p_5 + \frac{4}{3}up_3 + 2u_1p_2 + \Big(\frac{4}{9}u^2 + \frac{4}{3}u_2\Big)p_1 + \Big(\frac{4}{9}uu_1 + \frac{1}{3}u_3\Big)p_0 - \frac{1}{9}u_1w$$

with the corresponding operator

$$\mathscr{P}_3 = D_x^5 + \frac{4}{3}uD_x^3 + 2u_1D_x^2 + \Big(\frac{4}{9}u^2 + \frac{4}{3}u_2\Big)D_x + \Big(\frac{4}{9}uu_1 + \frac{1}{3}u_3\Big)p_0 - \frac{1}{9}u_1D_x^{-1}\circ u_1.$$

In CDE, let us first search for the operators. We have two possibilities: a weight approach and a CRACK approach. We start with the weight approach; see program file kdv_ho1.red. In order to produce an ansatz which is a superfunction of one odd variable (or a linear function in odd variables), we produce two lists, the list `l_grad_mon` of all even variables collected by their weights and a similar list `l_grad_odd` for odd variables:

```
l_grad_der:=der_deg_ordering(0,all_parametric_der);
l_grad_odd:=
  {1} . der_deg_ordering(1,all_parametric_odd);
gradmon:=graded_mon(1,10,l_grad_der);
gradmon:={1} . gradmon;
```

We need a list of weighted monomials which are linear in odd variables. The function `mkalllinodd` produces all monomials which are linear with respect to the variables from `l_grad_odd`, have (monomial) coefficients from the variables in `l_grad_der`, and have weights from 1 to 6. Such monomials are then converted to the internal representation of odd variables.

```
linodd:=mkalllinodd(gradmon,l_grad_odd,1,6);
```

Note that all odd variables have positive weights due to the assumption

```
deg_odd_var:={1}
```

Finally, the ansatz for local Hamiltonian operators:

```
sym:=(for each el in linext sum c(ctel:=ctel+1)*el);
```

We load the linearization as the $\mathscr{C}$-differential operator `lkdv` (see the example within Sect. 2.2.3). Note that the linearization must be restricted to the equation; on the other hand, the linearization is *automatically* lifted to the tangent covering.

After having set

```
equ 1:=lkdv(1,1,sym);
```

and having initialized the equation solver as before, we split the polynomial equation with respect to the odd variables

```
tel:=splitext_opequ(equ,1,1);
```

and split the resulting polynomial system in the system of the coefficients of all monomials:

```
tel2:=splitvars_opequ(equ,2,tel,vars);
```

Now we are ready to solve all equations:

```
put_equations_used tel;
for i:=2:tel do integrate_equation i;
```

The results are the two well-known Hamiltonian operators for the KdV:

```
sym := (c(5)*p*u_x + 2*c(5)*p_x*u + 3*c(5)*p_3x
  + 3*c(2)*p_x)/3;
```

Of course, the results correspond to the operators:

$$p_x \to D_x,$$
$$\frac{1}{3}(3p_{xxx} + 2up_x + u_x p) \to \frac{1}{3}(3D_{xxx} + 2uD_x + u_x).$$

Note that each operator is multiplied by one arbitrary real constant, `c(5)` and `c(2)`.

The same problem can be approached using CRACK, as follows (program file kdv_ho2.red). An ansatz is constructed by the following instructions:

```
even_vars:=for i:=0:3 join
  selectvars(0,i,dep_var,all_parametric_der);
odd_vars:=for i:=0:3 join
  selectvars(1,i,odd_var,all_parametric_odd);

ctemp:=0;
ansatz:=for each el in odd_vars sum
  mkid(s,ctemp:=ctemp+1)*el;
```

Note that we have

```
ansatz := p*s1 + p_2x*s3 + p_3x*s4 + p_x*s2;
```

Indeed, we are looking for a third-order operator whose coefficients depend on even variables of order not higher than 3. This last property has to be introduced by

```
unk:=for i:=1:ctemp collect mkid(s,i);
for each ell in unk do
 for each el in even_vars do depend ell,el;
```

Then, we introduce the linearization as above and compute the equation to be solved:

```
total_eq:=lkdv(1,1,ansatz);
```

Finally, we split the above equation by collecting all coefficients of odd variables:

```
system_eq:=splitext_list({total_eq});
```

and we feed CRACK with the overdetermined system of linear partial differential equations that consist in asking that the above coefficients be zero:

```
load_package crack;
crack_results:=crack(system_eq,{},unk,
   cde_difflist(all_parametric_der,even_vars));
```

The results are the same as in the previous computation:

```
crack_results := {{{},
{s4=(3*c_17)/2,s3=0,s2=c_16 + c_17*u,s1=(c_17*u_x)/2},
{c_17,c_16},
{}}};
```

Now, we should compute Schouten brackets of the operators to be sure that they are Hamiltonian operators and not just bivectors. Strictly speaking, this is not needed for KdV, as it fulfills the conditions stated in [60] for the vanishing of all three vectors. However, let us show how these computations work. We describe the program file kdv_ho3.red. After having defined the operators

```
mk_cdiffop(ham1,1,{1},1);
for all psi1 let ham1(psi1)=td(psi1,x);
mk_cdiffop(ham2,1,{1},1);
for all psi2 let ham2(psi2)=(1/3)*u_x*psi2
 + td(psi2,x,3) + (2/3)*u*td(psi2,x);
```

We may convert the two operators into the corresponding superfunctions:

```
conv_cdiff2superfun(ham1,sym1);
conv_cdiff2superfun(ham2,sym2);
```

The result of the conversion is

```
sym1(1) := {p_x};
sym2(2) := {(1/3)*p*u_x + p_3x + (2/3)*p_x*u};
```

Skew-adjointness is checked at once:

```
adjoint_cdiffop(ham1,ham1_star);
adjoint_cdiffop(ham2,ham2_star);
ham1_star_sf(1)+sym1(1);
ham2_star_sf(1)+sym2(1);
```

and the result of the last two commands is zero.

Then we shall convert the two superfunctions into bivectors (10.13):

```
conv_genfun2biv(sym1,biv1);
conv_genfun2biv(sym2,biv2);
```

The output is

```
biv1(1);

 - ext(p,p_x);

biv2(1);

(1/3)*( - 3*ext(p,p_3x) - 2*ext(p,p_x)*u);
```

Finally, the three Schouten brackets $[\![\mathscr{P}_i, \mathscr{P}_j]\!]$ are computed, with $i, j = 1, 2$. We can do this step by step, starting by variational derivatives.

```
pvar_df(0,biv1(1),u);

0

pvar_df(1,biv1(1),p);

 - 2*p_x

pvar_df(0,biv2(1),u);

  - 2*ext(p,p_x)
-----------------
        3

pvar_df(1,biv2(1),p);

 2*( - p*u_x - 3*p_3x - 2*p_x*u)
----------------------------------
                3
```

Then we compute the formula for Schouten brackets, which gives the answer up to total derivatives:

```
schouten_bracket(biv1,biv1,thr11);

thr11(1);

0

schouten_bracket(biv1,biv2,thr12);
```

```
thr12(1);

0

schouten_bracket(biv2,biv2,thr22);

thr22(1);

 8*ext(p,p_x,p_3x)
-------------------
         3
```

We can check that the last expression is a total derivative by computing its Euler operator:

```
euler_df(thr22(1));

{{0},{0}}
```

For simplicity, the following command directly checks if the Schouten bracket of two bivectors is zero:

```
iszero_schouten_bracket(biv1,biv1,th11);
```

The content of `th11(1)` is in this case

```
{{0},{0}};
```

A nonlocal Hamiltonian operator for the KdV equation is computed in the program file kdv_ho4.red as follows. We assume that we generated the cotangent covering with the nonlocal variable (9.3); see Sect. 9.2.4. Then we can search a nonlocal Hamiltonian operator using the weight approach. In particular, if we choose an ansatz of weight 7:

```
linodd:=mkalllinodd(gradmon,l_grad_odd,7,7);
phi:=(for each el in linodd sum (c(ctel:=ctel+1)*el));
```

we can solve the equation for shadows of symmetries

```
equ 1:=td(phi,t)-u*td(phi,x)-u_x*phi-td(phi,x,3);
```

with the usual splitting technique. We obtain the well-known first nonlocal Hamiltonian operator for the KdV equation:

```
phi := (c(1)*(4*p*u*u_x + 3*p*u_3x + 18*p_2x*u_x
 + 12*p_3x*u + 9*p_5x + 4*p_x*u**2 + 12*p_x*u_2x
 - r1*u_x))/4;
```

Note that the choice of the weight 7 is motivated by the fact that the first two (local) Hamiltonian operators have, respectively, weight 3 and 5, and the recursion operator has weight 2. The above nonlocal operator is, indeed, the result of the composition of the recursion operator with the second (local) Hamiltonian operator.

10.2.2 Dispersionless Boussinesq System

Let $\Phi = (\Phi^w, \Phi^u, \Phi^v)$ satisfy

$$\begin{aligned} D_t(\Phi^w) &= D_x(\Phi^u), \\ D_t(\Phi^u) &= wD_x(\Phi^w) + w_1\Phi^w + D_x(\Phi^v), \\ D_t(\Phi^v) &= -uD_x(\Phi^w) - 3u_1\Phi^w - 3wD_x(\Phi^u) - w_1\Phi^u. \end{aligned}$$

There exist three independent solutions Φ_1, Φ_2, and Φ_3 to this system with the components

$$\Phi_1^w = p_1^v, \quad \Phi_1^u = p_1^u, \quad \Phi_1^v = -4p_1^v w + p_1^w - 2p^v w,$$

$$\begin{aligned} \Phi_2^w &= 4p_1^v v + 3p_1^u u + 2p_1^w w + 2p^v v_1 + p^u u_1, \\ \Phi_2^u &= -11p_1^v wu + 2p_1^u(w^2 + 2v) + 3p_1^w u - 2p^v(wu_1 + 4uw_1) + p^u(ww_1 + v_1), \\ \Phi_2^v &= -p_1^v(6w^3 + 16wv + 3u^2) - 11p_1^u wu + 4p_1^w v \\ &\quad - 2p^v(3w^2w_1 + 2wv_1 + uu_1 + 4vw_1) - p^u(3wu_1 + uw_1), \end{aligned}$$

and

$$\begin{aligned} \Phi_3^w &= p^v v_1 + p^u u_1 + p^w w_1, \\ \Phi_3^u &= p^v(3wu_1 + uw_1) + p^u(ww_1 + v_1) + p^w u_1, \\ \Phi_3^v &= p^v(-3w_2w_1 - 4wv_1 - uu_1) + p^u(-3wu_1 - uw_1) + p^w v_1. \end{aligned}$$

The operator

$$\mathscr{P}_1 = \begin{pmatrix} 0 & 0 & D_x \\ 0 & D_x & 0 \\ D_x & 0 & -4wD_x - 2w_1 \end{pmatrix}$$

corresponding to the first solution is skew-adjoint and, as it is easily checked, is a Poisson structure. But to Φ_2 and Φ_3, there correspond the operators

$$\mathscr{P}_2 = \begin{pmatrix} 2wD_x & 3uD_x + u_1 & 2(2vD_x + v_1) \\ 3uD_x & 2(w^2 + 2v)D_x + ww_1 + v_1 & -11wuD_x - 2(wu_1 + 4uw_1) \\ 4vD_x & -11wuD_x - 3wu_1 - uw_1 & h_1D_x + h_0 \end{pmatrix},$$

where

$$h_1 = -(6w^3 + 16wv + 3u^2), \qquad h_0 = -2(3w^2w_1 + 2wv_1 + uu_1 + 4vw_1),$$

and

$$\mathscr{P}_3 = \begin{pmatrix} w_1 & u_1 & v_1 \\ u_1 & ww_1 + v_1 & -3wu_1 - uw_1 \\ v_1 & -3wu_1 - uw_1 & -3w^2w_1 - 4wv_1 - uu_1 \end{pmatrix}$$

which are not skew-adjoint. But their sum $\mathscr{P} = \mathscr{P}_2 + \mathscr{P}_3$ is a Poisson structure compatible with $\mathscr{P}_1$.

In CDE, after generating the cotangent covering as in Sect. 9.2.2, we can search for Hamiltonian operators in a way that is technically very similar to what was done for recursion and symplectic operators. Indeed, we will find Hamiltonian operators by a weight approach. In the program file bou_ho1.red, after defining weights

```
deg_indep_var:={-1,-2};
deg_dep_var:={3,4,2};
deg_odd_var:={2,1,3};
cde_grading(deg_indep_var,deg_dep_var,deg_odd_var);
```

We can repeat line by line the code that generates the ansatz as a graded polynomial. Then we set up the equation:

```
for i:=1:nc do
  equ(i):=for j:=1:nc sum lbou(i,j,phi(j));
```

where `lbou` is the linearization (lifted to the tangent covering). The solution of the above system yields three generating functions of bivectors:

```
phi(1) := ( - 7*c(59)*p*v_x - 7*c(59)*p*w*w_x
 + 4*c(59)*p_x*v + 2*c(59)*p_x*w**2
 + 22*c(59)*q*u_x*w - 11*c(59)*q_x*u*w
 - 8*c(59)*r*u_x + 3*c(59)*r_x*u - c(56)*p*v_x
 - c(56)*p*w*w_x - 12*c(56)*p_x*v - 6*c(56)*p_x*w**2
 + 22*c(56)*q*u*w_x + 33*c(56)*q_x*u*w
 + 2*c(56)*r*u_x - 9*c(56)*r_x*u + 22*c(6)*p_x)/22;
phi(2) := (7*c(59)*p*u*w_x + 21*c(59)*p*u_x*w
 - 11*c(59)*p_x*u*w + 6*c(59)*q*u*u_x
 - 8*c(59)*q*v*w_x + 28*c(59)*q*v_x*w
 + 18*c(59)*q*w**2*w_x - 3*c(59)*q_x*u**2
 - 16*c(59)*q_x*v*w - 6*c(59)*q_x*w**3
 - 8*c(59)*r*v_x + 4*c(59)*r_x*v + c(56)*p*u*w_x
 + 3*c(56)*p*u_x*w + 33*c(56)*p_x*u*w
 + 4*c(56)*q*u*u_x + 24*c(56)*q*v*w_x
 + 4*c(56)*q*v_x*w + 12*c(56)*q*w**2*w_x
 + 9*c(56)*q_x*u**2 + 48*c(56)*q_x*v*w
 + 18*c(56)*q_x*w**3 + 2*c(56)*r*v_x
 - 12*c(56)*r_x*v - 44*c(6)*q*w_x
 - 88*c(6)*q_x*w + 22*c(6)*r_x)/22;
phi(3) := ( - 7*c(59)*p*u_x + 3*c(59)*p_x*u
```

```
  - 6*c(59)*q*v_x + 4*c(59)*q_x*v - 8*c(59)*r*w_x
  + 2*c(59)*r_x*w - c(56)*p*u_x - 9*c(56)*p_x*u
  - 4*c(56)*q*v_x - 12*c(56)*q_x*v + 2*c(56)*r*w_x
  - 6*c(56)*r_x*w + 22*c(6)*q_x)/22;
```

We shall test for self-adjointness each of the three generating functions. The following code will do the job:

```
mk_superfun(ham1_sf,1,3);
for i:=1:nc do
  ham1_sf(i):=df(phi(i),c(6));
conv_superfun2cdiff(ham1_sf,ham1);
adjoint_cdiffop(ham1,ham1_star);
for i:=1:nc do if ham1_star_sf(i) + ham1_sf(i) neq 0
  then write "Non self-adjoint operator";
```

We shall repeat the code for the other two generating functions. It turns out that exactly the other two generating functions, selected by the constants `c(56)` and `c(59)`, are not skew-adjoint. However, the following code helps to find a solution within the vector space generated by the linear combination of the above two generating functions:

```
mk_superfun(ham0_sf,1,3);
for i:=1:nc do
  ham0_sf(i):=alpha*ham2_sf(i) + beta*ham3_sf(i);
conv_superfun2cdiff(ham0_sf,ham0);
adjoint_cdiffop(ham0,ham0_star);
for i:=1:nc do write ham0_star_sf(i) + ham0_sf(i);
```

The last equation shows clearly that the linear combination is a skew-adjoint generating function only for `alpha:=9` and `beta:=5`.

The Hamiltonian property is checked by first converting the skew-adjoint generating functions into bivectors

```
conv_genfun2biv(ham0_sf,biv0);
conv_genfun2biv(ham1_sf,biv1);
```

and then computing formula (10.6)

```
iszero_schouten_bracket(biv1,biv1,th11);
iszero_schouten_bracket(biv1,biv0,th10);
iszero_schouten_bracket(biv0,biv0,th00);
```

The results are three lists of zeros.

10.2.3 Camassa-Holm Equation

The defining equation

$$\alpha D_t(\Phi) - D_x^2 D_t(\Phi) - u D_x^3(\Phi) - 2u_x D_x^2(\Phi) + (3\alpha u - 2u_{xx}) D_x(\Phi) + (3\alpha u_x - u_{xxx})\Phi = 0$$

possesses two independent solutions

$$\Phi_1 = p_0^1, \qquad \Phi_2 = -p_1^0 - u p_0^1 + u_x p$$

with the corresponding operators

$$\mathscr{P}_1 = D_x, \qquad \mathscr{P}_2 = -D_t - u D_x + u_x,$$

which satisfy Condition 1. The simplest way to check Condition 2 is, probably, to rewrite the Camassa-Holm equation in the three-component evolutionary form:

$$u_x = v,$$
$$v_x = \alpha u - w,$$
$$w_x = -\frac{w_t + 2vw}{u}$$

and find the corresponding representations for $\mathscr{P}_1$ and $\mathscr{P}_2$.

In CDE, after loading the cotangent covering with one odd variable and all weights (Sect. 9.2.3), we solve the equation

```
equ 1:=num lch(1,1,psi);
```

where lch is the linearization of the Camassa-Holm equation lifted to the cotangent covering. The computation can be found in the program file ch_ho1.red. A weight approach leads to the above solutions:

```
psi:=(c(14)*alpha**2*p_x + c(11)*alpha*p*u*u_x
  - c(11)*alpha*p*u_t - c(11)*alpha*p_t*u
  + c(11)*alpha*p_x*u**2 - c(11)*o1*u_x
  - c(11)*p*u_2x*u_x + c(11)*p_2tx + c(11)*p_2x*u_t
  + c(11)*p_t2x*u - c(11)*p_x*u*u_2x
  + c(6)*alpha*p*u_x - c(6)*alpha*p_t
  - c(6)*alpha*p_x*u + c(3)*alpha*p_x)/alpha;
```

(here one solution is repeated two times).

10.2.4 WDVV System

As it was already pointed out, the simplest nontrivial case of the WDVV equations is the third-order Monge-Ampère equation, $f_{ttt} = f_{xxt}^2 - f_{xxx}f_{xtt}$, [28]. This PDE is a differential covering of the following quasilinear first-order system of PDEs

$$a_t = b_x, \quad b_t = c_x, \quad c_t = (b^2 - ac)_x,$$

via the map $a = f_{xxx}$, $b = f_{xxt}$, $c = f_{xtt}$. This system possesses two Hamiltonian operators, [31]:

$$\begin{pmatrix} a \\ b \\ c \end{pmatrix}_t = A_i \begin{pmatrix} \delta H_i/\delta a \\ \delta H_i/\delta b \\ \delta H_i/\delta c \end{pmatrix}, \qquad i = 1, 2,$$

with Hamiltonian densities H_i. The operators are a first-order Hamiltonian operator

$$A_1 = \begin{pmatrix} -\frac{3}{2}D_x & \frac{1}{2}D_x a & D_x b \\ \frac{1}{2}aD_x & \frac{1}{2}(D_x b + bD_x) & \frac{3}{2}cD_x + c_x \\ bD_x & \frac{3}{2}D_x c - c_x & (b^2 - ac)D_x + D_x(b^2 - ac) \end{pmatrix}$$

and a third-order Hamiltonian operator

$$A_2 = D_x \begin{pmatrix} 0 & 0 & D_x \\ 0 & D_x & -D_x a \\ D_x & -aD_x & D_x b + bD_x + aD_x a \end{pmatrix} D_x.$$

The operator A_2 belongs to a class that has been defined in [27]. Such operators have been classified in [33, 34]. See also [35].

It is easy to input the two operators in CDE. We refer to the example file wdvvs_ho1.red. We load the jet space and the differential equation by

```
indep_var:={t,x};
dep_var:={a,b,c};
odd_var:={p,q,r};
total_order:=10;
principal_der:={a_t,b_t,c_t};
de:={b_x,c_x,(2*b*b_x - a*c_x - a_x*c)};
nc:=length(dep_var);
```

Let us start by A_1. Define `gu1` as a matrix that contains the coefficients:

$$g^{ij} = \begin{pmatrix} -\frac{3}{2} & \frac{1}{2}a & b \\ \frac{1}{2}a & b & \frac{3}{2}c \\ b & \frac{3}{2}c & 2(b^2 - ac) \end{pmatrix}$$

in Reduce:

```
gu1:=mat(
(-(3/2),(1/2)*a,b),
((1/2)*a,b,(3/2)*c),
(b,(3/2)*c,2*(b**2 - a*c)));
```

Then, define the matrix gamma_hi_con by:

```
gamma_hi_con:=mat(
(0,(1/2)*a_x,b_x),
(0,(1/2)*b_x,c_x),
(0,(1/2)*c_x,2*b*b_x - a_x*c - a*c_x)
);
```

Finally, define the operator A_1:

```
mk_cdiffop(aa1,1,{3},3);
for all i,j,psi let aa1(i,j,psi)=
  gu1(i,j)*td(psi,x) + gamma_hi_con(i,j)*psi;
```

The third order operator A_2 can be input as follows. Observe that the leading contravariant metric is

$$g^{ij} = \begin{pmatrix} 0 & 0 & 1 \\ 0 & 1 & -a \\ 1 & -a & 2b + a^2 \end{pmatrix}$$

Introduce the above matrix in Reduce as gu3. Then, introduce the matrix

```
c_hi_con:=mat(
(0,0,0),
(0,0, - a_x),
(0,0,b_x + a*a_x)
);
```

Finally, define the operator A_2:

```
mk_cdiffop(aa2,1,{3},3);
for all i,j,psi let aa2(i,j,psi) =
td(
  gu3(i,j)*td(psi,x,2)+c_hi_con(i,j)*td(psi,x)
    ,x
);
```

Now, we can test the Hamiltonian property of A_1 and A_2 and their compatibility:

```
conv_cdiff2genfun(aa1,sym1);
conv_cdiff2genfun(aa2,sym2);

conv_genfun2biv(sym1,biv1);
```

```
conv_genfun2biv(sym2,biv2);

iszero_schouten_bracket(biv1,biv1,sb11);
iszero_schouten_bracket(biv1,biv2,sb12);
iszero_schouten_bracket(biv2,biv2,sb22);
```

Needless to say, the result of the last three commands is a list of zeroes.

We observe that the same software can be used to prove the bi-Hamiltonianity of a *six*-component WDVV system, cf. [114].

10.2.5 Multi-dimensional Examples

Consider multi-dimensional systems.

10.2.5.1 The Kadomtsev-Petviashvili Equation

Let us work first in the non-evolutionary presentation, program file kp_ho1.red. We use the cotangent covering and weights as in Sect. 9.2.4. After making an ansatz with weights from 1 to 5:

```
linodd:=mkalllinodd(gradmon,l_grad_odd,1,5);
phi:=for each el in linodd sum c(ctel:=ctel+1)*el;
```

we set up the equation for shadows of symmetries in the cotangent covering

```
equ 1:=td(phi,y,2) - td(phi,x,t) + 2*u_x*td(phi,x)
 + u_2x*phi + u*td(phi,x,2) + (1/12)*td(phi,x,4);
```

and solve it. We obtain

```
p_2x;
```

which is the only known local Hamiltonian operator for the KP equation.

In the evolutionary presentation (program file kpev_ho1.red), using the cotangent covering and weights from Sect. 6.2.4, we generate an ansatz for the two components of the shadow of a symmetry on the cotangent covering that we are looking for:

```
linodd:=mkalllinodd(gradmon,l_grad_odd,1,5);
psi1:=for each el in linodd sum c(ctel:=ctel+1)*el;
psi2:=for each el in linodd sum c(ctel:=ctel+1)*el;
```

After loading the linearization

```
% Equation for shadows of symmetries in
% the cotangent covering
mk_superfun(lkpev_sf,1,2);
lkpev_sf(1):=p_y - q_x;
lkpev_sf(2):=(12*p*u_x + p_3x - 12*p_t + 12*p_x*u
  + 12*q_y)/12;
conv_superfun2cdiff(lkpev_sf,lkpev);
```

we can set up the equations:

```
equ 1:=lkpev(1,1,psi1) + lkpev(1,2,psi2);
equ 2:=lkpev(2,1,psi1) + lkpev(2,2,psi2);
```

We obtain the solution:

```
psi1 := q_x;
psi2 := p_x;
```

which corresponds to the operator

$$\begin{pmatrix} 0 & D_x \\ D_x & 0 \end{pmatrix}.$$

10.2.5.2 Three-Field Hierarchy

This is an example from [17]. We consider the equation

$$u_t = u_y + 2v_x, \quad v_t = uv_x + 2w_x, \quad w_t = u_x w + uw_x. \tag{10.17}$$

The above equation is shown to be integrable in the sense that it has a Hamiltonian operator and a hierarchy of commuting symmetries.

Here we find the Hamiltonian operator in our framework. We can compute the linearization and its adjoint by the program file tfh_ell1.red. The Hamiltonian operator is found by the program file tfh_ho1.red. In this last file, the jet space is

```
indep_var:={t,x,y};
dep_var:={u,v,w};
odd_var:={p,q,r};
total_order:=8;
```

and the cotangent covering of the three-field hierarchy:

```
principal_der:={u_t,v_t,w_t};
de:={ u_y + 2*v_x,u*v_x + 2*w_x,u_x*w + u*w_x};
principal_odd:={p_t,q_t,r_t};
de_odd:={
p_y - q*v_x + r_x*w,
2*p_x + q*u_x + q_x*u,
2*q_x + r_x*u};
```

The weights here are quite straightforward, and their effect is just taking ordinary polynomials for the ansatz:

```
deg_indep_var:={-1,-1,-1};
deg_dep_var:={1,1,1};
deg_odd_var:={1,1,1};
```

The space of monomials is generated by

```
% List of variables ordered by weights
l_grad_mon:=der_deg_ordering(0,all_parametric_der);
l_grad_odd:=
  {1} . der_deg_ordering(1,all_parametric_odd);
% List of weighted monomials of weight <= 10
gradmon:=graded_mon(1,5,l_grad_mon);
gradmon:={1} . gradmon;
% list of weighted monomials which are
% linear in odd variables
% and have weights from 1 to 3
linodd:=mkalllinodd(gradmon,l_grad_odd,1,3);
```

The ansatz for local Hamiltonian operators is defined by

```
operator phi;
for i:=1:nc do phi(i):=
  (for each el in linodd sum (c(ctel:=ctel+1)*el));
```

We load the linearization of the three-field hierarchy equation (computed in the program file tfh_ell1.red together with its adjoint):

```
mk_superfun(ltfh_sf,1,3);
ltfh_sf(1):=p_t - p_y - 2*q_x;
ltfh_sf(2):= - p*v_x + q_t - q_x*u - 2*r_x;
ltfh_sf(3):= - p*w_x - p_x*w - r*u_x + r_t - r_x*u;
conv_superfun2cdiff(ltfh_sf,ltfh);
```

Then, we define the equation for shadows of symmetries:

```
for i:=1:nc do
equ(i):=for j:=1:nc sum ltfh(i,j,phi(j));
```

Solving it in the usual way, we obtain the vector superfunction that corresponds to the operator (6.16) in [17]:

```
phi 1;

c(15)*r_x;

phi 2;

(c(15)*(2*q_x + r_x*u - r_y))/2;

phi 3;

(c(15)*(2*p_x + q*u_x + q_x*u - q_y))/2;
```

The operator is local, and in CDE it is easy to check that its Schouten bracket vanishes. First of all, we generate a superfunction with the Hamiltonian operator:

```
mk_superfun(ham1_sf,1,3);
for i:=1:nc do ham1_sf(i):=df(phi(i),c(15));
conv_superfun2cdiff(ham1_sf,ham1);
```

We must check that the operator is skew-adjoint:

```
adjoint_cdiffop(ham1,ham1_star);
for i:=1:nc do if ham1_star_sf(i) + ham1_sf(i) neq 0
  then write "Non self-adjoint operator";
```

Then, we convert the above operator into a bivector and check if its Schouten bracket is zero:

```
conv_genfun2biv(ham1_sf,biv1);
iszero_schouten_bracket(biv1,biv1,thr11);
{{0,0,0},{0,0,0}};
```

the result being the list of all even and odd variational derivatives of the three-vector `thr11`.

Remark 10.4 We observe that the above Hamiltonian operator is of Dubrovin-Novikov type [27]; such operators are classified in the *three*-component situation [32]. In particular, the above operator can be transformed into the first canonical form of Theorem 3 in [32].

10.2.5.3 The Plebanski Equation

Since the Plebanski equation is Lagrangian, its Hamiltonian operators are its recursion operators that satisfy the integrability conditions (see Sect. 7.2.3). Note that in the paper [105] several Hamiltonian structures are exhibited. We stress that we can obtain all of them by the change of coordinate formula (see [82]) on passing to an evolutionary presentation. In particular, the identity map, which is the trivial recursion operator, becomes a nontrivial local Hamiltonian operator after the change of coordinates, thus hiding its much more simple nature.

Chapter 11
Recursion Operators for Cosymmetries

Abstract These $\mathscr{C}$-differential operators are somewhat dual (see Remark 11.1 below) to recursion operators for symmetries considered in Sect. 7. They send cosymmetries of an equation $\mathscr{E}$ to themselves. (Actually, recursion operators for cosymmetries take solutions of the equation $\tilde{\ell}^*_{\mathscr{E}}(\psi) = 0$ in some covering over $\mathscr{E}$, i.e. shadows of cosymmetries, to objects of the same nature.) Though these operators are not so popular in applications as their symmetry counterparts, we expose briefly our approach to compute them. In this chapter we give the solution to Problems 1.27 and 1.28.

11.1 General Theory

Let $\mathscr{O}: \hat{P} = \mathscr{F}(\mathscr{E}; r) \to \hat{P} = \mathscr{F}(\mathscr{E}; r)$ be of the form $\mathscr{O} = (\mathscr{O}^j_i)$ with

$$\mathscr{O} = \sum_\sigma b^j_{i,\sigma} D_\sigma, \qquad i, j, = 1, \dots, r.$$

Using the already standard scheme, we start our constructions with the vector-function $\psi_{\mathscr{O}} = (\psi^1_{\mathscr{O}}, \dots, \psi^r_{\mathscr{O}})$ having the form

$$\psi^j_{\mathscr{O}} = \sum_{i,\sigma} b^j_{i,\sigma} p^i_\sigma$$

and defined on $\mathscr{T}^*\mathscr{E}$.

Let $\tilde{\ell}^*_{\mathscr{E}}$ be the lift of the linearization operator to $\mathscr{T}^*\mathscr{E}$. Consider the equation

$$\tilde{\ell}^*_{\mathscr{E}}(\psi_{\mathscr{O}}) = 0. \tag{11.1}$$

J. Krasil'shchik et al., *The Symbolic Computation of Integrability Structures for Partial Differential Equations*, Texts and Monographs in Symbolic Computation,
https://doi.org/10.1007/978-3-319-71655-8_11

Then, similar to the previous situations, solutions of (11.1) are in one-to-one correspondence with the classes of $\mathscr{C}$-differential operators

$$\frac{\{\,\mathscr{O} \mid \ell^*_{\mathscr{E}} \circ \mathscr{O} = \Delta \circ \ell^*_{\mathscr{E}}\,\}}{\{\,\mathscr{O} \mid \mathscr{O} = \nabla \circ \ell^*_{\mathscr{E}}\,\}}, \tag{11.2}$$

where $\Delta\colon \mathscr{F}(\mathscr{E}; m) = \hat{P} \to \mathscr{F}(\mathscr{E}; m) = \hat{P}$ and $\nabla\colon \mathscr{F}(\mathscr{E}; m) \to \mathscr{F}(\mathscr{E}; r)$ are some $\mathscr{C}$-differential operators. Due to Eq. (11.2), nontrivial solutions of Eq. (11.1) are identified with nontrivial operators acting from cosymmetries of $\mathscr{E}$ to cosymmetries. However, in practice recursion operators are almost never local[1] and consequently may take a local (co)symmetry to a shadow.

Remark 11.1 Take a solution $\psi_{\mathscr{O}}$ of Eq. (11.1). Then, due to the above said, one has

$$\ell^*_{\mathscr{E}} \circ \mathscr{O} = \Delta \circ \ell^*_{\mathscr{E}}.$$

Conjugating this equality, we obtain

$$\ell_{\mathscr{E}} \circ \Delta^* = \mathscr{O}^* \circ \ell_{\mathscr{E}}.$$

Hence, Δ^* is a recursion operator for symmetries; vice versa, conjugating the similar equation on recursion operators for symmetries, one obtains the ones for cosymmetries. In particular, if $\mathscr{E}$ is an evolution equation and $\mathscr{R}$ is a recursion operator for its symmetries, then $\mathscr{R}^*$ is a recursion operator for cosymmetries and vice versa. Nevertheless, recursion operators for cosymmetries may be computed directly as well.

Example 11.1 Consider the Burgers equation

$$u_t = uu_x + u_{xx}$$

and its cotangent covering defined by

$$p_t = up_x - p_{xx}.$$

To find a recursion operator for cosymmetries, one needs to solve the equation $\tilde{\ell}^*_{\mathscr{E}}(\psi) = 0$ on the cotangent covering with ψ linear in p_k. Here

$$\tilde{\ell}^*_{\mathscr{E}} = D_x^2 - u_x D_x - D_t.$$

[1] As we noted already, mathematical folklore says that an equation admits a local recursion operator only if it is linear with constant coefficients, but we never saw a rigorous and more or less general proof.

The only solution is $\psi = p$ that corresponds to the identical operator. Nevertheless, if we add the nonlocal variable w canonically associated with the symmetry u_x, we shall find another solution

$$\psi = p_{xx} + \frac{1}{2}up - \frac{1}{2}w,$$

to which the operator

$$\mathscr{O}_\psi = D_x^2 + \frac{1}{2}u - \frac{1}{2}D_x^{-1} \circ u_x$$

corresponds, but the action of this operator on the only cosymmetry of the Burgers equation is trivial: $\mathscr{O}_\psi(1) = 0$.

Example 11.2 For the sine-Gordon equation

$$u_{xy} = \sin u$$

the cotangent covering is given by

$$p_{xy} = \cos u \cdot p$$

and recursion operators for symmetries are obtained from solutions of the equation

$$\tilde{D}_{xy}(\psi) = \cos u \cdot \psi.$$

Adding the canonical nonlocal variable associated with u_x, we obtain the solution

$$\psi = p_{xx} + u^2 p + u_{xx} w$$

with the corresponding recursion operator

$$\mathscr{O}_\psi = D_x^2 + u^2 + u_{xx} D_x^{-1} \circ u_x.$$

In a similar way, if we introduce the nonlocal variable associated with u_y, we shall get the operator

$$\mathscr{O}_\psi = D_y^2 + u^2 + u_{yy} D_y^{-1} \circ u_y.$$

Example 11.3 Consider the Harry Dym equation

$$u_t = u^3 u_{xxx}.$$

Its cotangent covering was presented in Example 10.3. To find recursion operators for cosymmetries, we need to solve the equation

$$\tilde{D}_t(\psi) = (6u_x^3 + 18uu_xu_{xx})\psi + (18uu_x^2 + 9u^2u_{xx})\tilde{D}_x(\psi) + 9u^2u_x\tilde{D}_x^2(\psi) + u^3\tilde{D}_x^3(\psi)$$

on $\mathcal{T}^*\mathcal{E}$. Extending the cotangent covering with the canonical nonlocal variable w associated with the symmetry u_t, we get the solution

$$\psi = u^2 p_{xx} + 5uu_x p_x + (4uu_{xx} + 3u_x^2)p - \frac{1}{u^2}w,$$

to which the recursion operator

$$\mathcal{O}_\psi = u^2 D_x^2 + 5uu_x D_x + (4uu_{xx} + 3u_x^2) - \frac{1}{u^2}D_x^{-1} \circ u^3 u_{xxx}$$

corresponds.

11.2 Examples

Let us pass to the computer implementation.

11.2.1 Korteweg-de Vries Equation

The defining equation reads

$$D_t(\psi) = uD_x(\psi) + D_x^3(\psi),$$

where the total derivatives are taken on $\mathcal{T}^*\mathcal{E}$ and $\psi = \sum_k b_k p_k$. The equation admits the trivial solution only. Extending the cotangent covering space with the nonlocal variable w defined by $w_x = u_x p$, we find the solution

$$\psi = p_2 + \frac{2}{3}up_0 - \frac{1}{3}w,$$

to which the recursion operator

$$\mathcal{O} = D_x^2 + \frac{2}{3}u - \frac{1}{3}D_x^{-1} \circ u_x \tag{11.3}$$

corresponds.

In CDE, we generate the jet space with the above nonlocal odd variable (program file kdv_roc1.red) as follows:

```
indep_var:={x,t};
dep_var:={u};
odd_var:={p,o};
total_order:=10;
```

The cotangent covering is loaded by

```
principal_odd:={p_t,o_x,o_t};
de_odd:={u*p_x+p_3x,
  p*u_x,
p*u*u_x + p*u_3x + p_2x*u_x - p_x*u_2x};
```

After loading the adjoint operator

```
mk_superfun(lkdv_star_sf,1,1);
lkdv_star_sf(1):=p_t - u*p_x - p_3x;
conv_superfun2cdiff(lkdv_star_sf,lkdv_star);
```

we need to solve the equation

```
equ 1:=lkdv_star(1,1,sym);
```

where sym is an appropriate ansatz, linear with respect to odd variables. We can use either the weights or a CRACK approach. Consistent weights are

```
deg_indep_var:={-1,-3};
deg_dep_var:={2};
deg_odd_var:={1,3};
```

An ansatz which depends on second-order odd variables produces the solutions

```
sym := ( - c(3)*o + 2*c(3)*p*u + 3*c(3)*p_2x
  + 2*c(1)*p)/2;
```

where the nontrivial solution is (11.3).

11.2.2 Dispersionless Boussinesq System

We must generate the cotangent covering with at least some nonlocal odd variables in order to obtain a first nontrivial solution. We describe the program file bou_roc1.red (the calculations are taken from [61]):

```
indep_var:={x,t};
dep_var:={u,v,w};
odd_var:={p,q,r,o1,o2,o3};
total_order:=8;
```

We also load a symmetry that will be used to define a canonical conservation law on the cotangent covering:

```
s3u:=w_x*w + v_x;
s3v:= - w_x*u - 3*u_x*w;
s3w:=u_x;
```

The odd equations are

```
principal_odd:={p_t,q_t,r_t,
  o1_t,o1_x,
  o2_t,o2_x,
  o3_t,o3_x};
de_odd:={
   - 2*q*w_x - 3*q_x*w + r_x,
  p_x,
  p_x*w + 2*q*u_x - q_x*u,
  p,q,
  r*u_x + p*(v_x+w_x*w) - q*(3*u_x*w + w_x*u),
  p*u_x + q*v_x + r*w_x,
  r*s3u + p*(s3v + s3w*w) - q*(3*s3u*w + s3w*u),
  p*s3u + q*s3v + r*s3w
  };
```

The weight approach goes as in the case of recursion operators (Sect. 7.2.2). We just note that consistent weights are given by

```
deg_indep_var:={-1,-2};
deg_dep_var:={3,4,2};
deg_odd_var:={2,1,3,1,5,6};
```

After loading the adjoint linearization

```
mk_superfun(lbou_star_sf,1,3);
lbou_star_sf(1):= - p_t - 2*q*w_x - 3*q_x*w + r_x;
lbou_star_sf(2):=p_x - q_t;
lbou_star_sf(3):=p_x*w + 2*q*u_x - q_x*u - r_t;
conv_superfun2cdiff(lbou_star_sf,lbou_star);
```

we can define the equation

```
for i:=1:nc do
  equ(i):=for j:=1:nc sum lbou_star(i,j,phi(j));
```

where `phi(j)` has been previously defined as weighted polynomials which are linear in odd variables. The usual algebraic solving of the above equations yields the solutions

```
phi(1)  := ( - c(39)*o3 - 4*c(39)*p*v
  - 2*c(39)*p*w**2 + 11*c(39)*q*u*w - 3*c(39)*r*u
  + 11*c(3)*p)/11;
phi(2)  := ( - 2*c(39)*o2 - 3*c(39)*p*u - 4*c(39)*q*v
 - 2*c(39)*r*w + 11*c(3)*q)/11;
phi(3)  := ( - 4*c(39)*o2*w - c(39)*p*u*w
 + 3*c(39)*q*u**2 + 6*c(39)*q*w**3 - 4*c(39)*r*v
 - 8*c(39)*r*w**2 + 11*c(3)*r)/11;
```

11.2.3 Camassa-Holm Equation

We can load the cotangent covering with one nonlocal odd variable as in Sect. 9.2.3. Then we can search recursion operators for cosymmetries using again the weight approach. We describe the program file ch_roc1.red. After loading the adjoint linearization `lch_star`, we set up the equation

```
equ 1:=num lch_star(1,1,psi);
```

on a previously generated ansatz. The algebraic solving yields a nontrivial and nonlocal recursion operator for cosymmetries:

```
psi := (2*c(4)*alpha**2*p + 2*c(2)*alpha*p*u
 - c(2)*o1 - c(2)*p*u_2x - c(2)*p_2x*u - c(2)*p_tx
 + 2*c(1)*alpha*p)/(2*alpha);
```

11.2.4 Multi-dimensional Examples

In view of Remark 11.1, we will just consider one example of multi-dimensional system.

11.2.4.1 Husain Heavenly Equation

The Husain heavenly equation

$$v_{ty}v_{pz} - v_{tz}v_{py} + v_{tt} + v_{pp} = 0 \tag{11.4}$$

was introduced in [47]. In [136], it is shown that the equation has a *partner symmetry*. We could provide a recursion operator as we did for the Plebanski equation in Sect. 7.2.5.1 and observe that (11.4) is a Lagrangian equation, i.e., $\ell_{\mathcal{E}} = \ell^*_{\mathcal{E}}$; in this case all types of IS coincide.

However, we can show that our method works also in the multi-dimensional situation using an evolutionary presentation of the Husain heavenly equation. Namely, we will use the Husain system from [136]:

$$v_t = q, \qquad q_t = q_z v_{py} - q_y v_{pz} - v_{pp}, \tag{11.5}$$

with the aim of finding a recursion operator for cosymmetries C. This is the adjoint of the recursion operator for symmetries that is presented in the same paper, Eq. (9.6):

$$C = \begin{pmatrix} (v_{pz}D_y - v_{py}D_z) \circ D_p^{-1} & q_zD_y - q_yD_z - D_p \\ D_p^{-1} & 0 \end{pmatrix} \tag{11.6}$$

We compute the linearization of the above system and its adjoint in the program file hh_ell1.red. Then, we need to find a nonlocal variable that will reproduce the above operator C. To this aim we search for a conservation law ω on the cotangent covering which is linear and has two components. Given the above operator, we make the ansatz

$$\omega = P dp \wedge dy \wedge dz + T dt \wedge dy \wedge dz \tag{11.7}$$

and solve the problem by the program file hh_nlv1.red. Here, we load the jet space

```
indep_var:={t,p,y,z};
dep_var:={v,q};
odd_var:={w1,w2};
total_order:=6;
```

the even equation

```
principal_der:={v_t,q_t};
de:={q,q_z*v_py - q_y*v_pz - v_2p};
```

and the odd equation

```
ctc_hh:={
q_py*w2_z - q_pz*w2_y + q_y*w2_pz
- q_z*w2_py - w1_t + w2_2p,
v_py*w2_z - v_pz*w2_y - w1 - w2_t};

cotan_covering_eq:=first
  solve({part(ctc_hh,1),part(ctc_hh,2)},
    {w1_t,w2_t});

principal_odd:={w1_t,w2_t};
de_odd:={rhs(first cotan_covering_eq),
  rhs(second cotan_covering_eq)};
```

After calling CDE, we need an ansatz for the conservation law coefficients. After initialization

```
cnt:=0;
unk_cf:={};
nc:=length(dep_var);
```

we built an ansatz which is linear in odd variables (up to the 3rd order) and has coefficient functions of unspecified form and which depends upon 3rd- order derivatives of even variables.

```
even_vars:=for i:=0:3 join
  selectvars(0,i,dep_var,all_parametric_der);
odd_vars:=for i:=0:3 join
  selectvars(1,i,odd_var,all_parametric_odd);
```

```
ansatz:=for each el in odd_vars sum
  <<
    unk_cf:=mkid(cf_,cnt:=cnt+1) . unk_cf;
    mkid(cf_,cnt)*el
  >>;
for each el in unk_cf do
  for each ell in even_vars do depend el,ell;
```

Note that we collected all unknown coefficient functions in the list `unk_cf`. The ansatz for the conservation law is

```
cp:=w1;
ct:=ansatz;
```

Note that we assumed the component P (represented in the program as `cp`) to be equal to w^1, the first odd variable in the cotangent covering. The equation is

```
total_eq:=td(cp,t) - td(ct,p);
```

and, before solving, it is necessary to isolate the coefficients of the odd variables:

```
system_eq:=splitext_list({total_eq});
```

We solve the system with the help of `CRACK`:

```
load_package crack;
crack_results:=crack(system_eq,{},unk_cf,
   cde_difflist(all_parametric_der,even_vars));
```

The result is

```
{{{},
{cf_1=0,cf_2=0,cf_3=0,cf_4=1,cf_5=0,cf_6= - q_z,
cf_7=0,cf_8=q_y,cf_9=0,cf_10=0,cf_11=0,cf_12=0,
cf_13=0,cf_14=0,cf_15=0,cf_16=0,cf_17=0,cf_18=0,
cf_19=0,cf_20=0,cf_21=0,cf_22=0,cf_23=0,cf_24=0,
cf_25=0,cf_26=0,cf_27=0,cf_28=0,cf_29=0,cf_30=0,
cf_31=0,cf_32=0,cf_33=0,cf_34=0,cf_35=0,cf_36=0,
cf_37=0,cf_38=0,cf_39=0,cf_40=0},{},
{}}};
```

Replacing this solution into the ansatz yields

```
sub(second first crack_results,ansatz);


q_y*w2_z - q_z*w2_y + w2_p;
```

which is the second component of the conservation law.

Now, we search for the recursion operator for cosymmetries; we use the program file hh_roc1.red. After loading the cotangent covering, we build a second-order ansatz using the following monomials:

```
even_vars:=for i:=0:2 join
  selectvars(0,i,dep_var,all_parametric_der);
all_vars:=1 . even_vars;
odd_vars:=for i:=0:2 join
  selectvars(1,i,odd_var,all_parametric_odd);
ansatz_mon:=
  for each el in odd_vars join
    for each ell in all_vars collect ell*el;
```

The ansatz for the operator C is

```
operator phi;
phi(1):=(for each el in ansatz_mon sum
   (c(ctel:=ctel+1)*el));
phi(2):=(for each el in ansatz_mon sum
   (c(ctel:=ctel+1)*el));
```

We load the adjoint linearization operator:

```
mk_superfun(lhh_star_sf,1,2);
lhh_star_sf(1):=q_py*w2_z - q_pz*w2_y + q_y*w2_pz
   - q_z*w2_py - w1_t + w2_2p;
lhh_star_sf(2):=v_py*w2_z - v_pz*w2_y - w1 - w2_t;

conv_superfun2cdiff(lhh_star_sf,lhh_star);
```

The equations for shadows of cosymmetries on the cotangent covering are defined by

```
for i:=1:nc do
  equ(i):=num(for j:=1:nc sum lhh_star(i,j,phi(j)));
```

The solution is

```
phi 1;

c(85)*q_y*w2_z - c(85)*q_z*w2_y + c(85)*r1_y*v_pz
- c(85)*r1_z*v_py + c(85)*w2_p + c(1)*w1;

phi 2;

 - c(85)*r1 + c(1)*w2;
```

Note the presence of the trivial solution that corresponds to `c(1)`. Evidently, the solution that corresponds to `c(85)` is the operator C.

Discussion

The theoretical setting and the software described in this book provide facilities for the computation of important invariants of differential equations responsible for their integrability, including

- symmetries,
- cosymmetries and conservation laws,
- recursion operators (for both symmetries and cosymmetries),
- symplectic structures,
- Poisson structures (Hamiltonian operators).

However, there are some aspects and areas in this field that are not covered by either the theory or the software (or both) and need further work not only on program implementation, but also on theoretical research. Let us list those ones that seem to us to be most important and interesting at the moment.

- The current version of `CDE` can check integrability properties (Nijenhuis bracket, Schouten bracket and super-Euler operator) for local structures only (see Chaps. 7, 8, and 10). In practice, the arising structures are often (and almost always in the case of recursion operators) nonlocal. Theoretical foundations of nonlocal integrability are also rather weak. So, we need a good theory and efficient computer implementation for nonlocal IS.
- The computational scheme for nonlocal IS adopted in this book works as follows: we introduce conservation laws on tangent and cotangent coverings that are canonically associated with cosymmetries and symmetries, respectively, of the equation $\mathcal{E}$. This construction is developed for weakly nonlocal operators only. It would be interesting and useful to implement methods that produce nonlocal operators of a more general nature. This task is also related to the next two ones.
- We have a well-elaborate method to construct Abelian coverings, i.e., the ones that are associated with two-component conservation laws, but non-Abelian coverings also play an important role in geometry of PDEs and integrable

J. Krasil'shchik et al., *The Symbolic Computation of Integrability Structures for Partial Differential Equations*, Texts and Monographs in Symbolic Computation,
https://doi.org/10.1007/978-3-319-71655-8

systems. Construction of more general coverings may base on computation of *Wahlquist-Estabrook algebras* (see [143, 144] and [85]) and their representations as vector fields on $\mathbb{R}^n$.

- As we saw in Sect. 7.1.3, a recursion operator for symmetries may be interpreted as an auto-Bäcklund transformation of the tangent covering (and, naturally, those of the cotangent covering are recursion operators for symmetries). A natural step is to consider Bäcklund transformations between $\mathcal{T}\mathcal{E}$ and $\mathcal{T}^*\mathcal{E}$. They generalize and glue together the notions of symplectic and Poisson structures and are in relation to *Dirac structures* by I. Dorfman [24].
- Lax pairs and zero curvature representations (ZCR) (e.g., see [95] and references therein) are particular types of coverings also essential for many integrability problems, and computer-based procedures that provide their construction would be helpful for researchers.
- Especially important are those Lax pairs and ZCR that contain a spectral parameter. For example, Bäcklund transformations that originate from coverings with a parameter often possess the *permutability property*. But the parameter must be essential, i.e., nonremovable by an equivalence transformation. The problem of a parameter (non)removability is to be solved for practical examples, cf. [98].
- Finally, there exists a set of problems related to symmetry reductions of PDEs. They include:
  - description of finite-dimensional (solvable) subalgebras in $\operatorname{sym}\mathcal{E}$,
  - their classification up to conjugation,
  - construction of reductions corresponding to the found subalgebras.

We hope that all these problems will be solved step-by-step and their implementations will be included in future releases of `CDE`. We also believe that new problems will arise during the practical work with the system.

Acknowledgments

We would like to thank our friend and colleague Paul H.M. Kersten, who cooperated with us in using and understanding his package `CDIFF` for many years.

RV would like to thank A.C. Norman for his help and support in understanding Reduce internals.

Special thanks are also due to A. Sergyeyev and F.J. Wright for carefully reading and commenting parts of the book.

We also would like to thank all participants to the Seminar on the Geometry of Differential Equations, run by one of us (ISK) at the Independent University of Moscow. See http://gdeq.org for more details.

RV would also like to thank H. Baran, M. Casati, E.V. Ferapontov, A. Fordy, B.G. Konopelchenko, P. Lorenzoni, G. Manno, L. Martina, F. Oliveri, G. Saccomandi, M.V. Pavlov, and P. Vojčák for many interesting discussions and computations.

Thanks are due to A. Falconieri, the system administrator of the workstation `sophus` of the Dipartimento di Matematica e Fisica "E. De Giorgi," where many of the computations presented in this book have been run.

The work of ISK was partially supported by the *Simons-IUM fellowship* and *2017 Dobrushin professor* grant.

The work of ISK and RV was also supported by the project "More Electric Aircraft," Dipartimento di Ingegneria dell'Innovazione, Università del Salento, Lecce, Italy.

The work of ISK, AMV, and RV is partially supported by GNFM of Istituto Nazionale di Alta Matematica, by Istituto Nazionale di Fisica Nucleare under the IS-CSN4 *Mathematical Methods of Nonlinear Physics*, and by the Dipartimento di Matematica e Fisica dell'Università del Salento.

J. Krasil'shchik et al., *The Symbolic Computation of Integrability Structures for Partial Differential Equations*, Texts and Monographs in Symbolic Computation,
https://doi.org/10.1007/978-3-319-71655-8

Appendix A
Getting Started with Reduce and CDE

A.1 Installation of Reduce

In order to use the CDE package, it is enough to have a recent version of Reduce with both the CDE and the CDIFF packages installed. This section is aimed at explaining how to install Reduce.

Nearly all information on Reduce can be found at the Reduce web page on SourceForge [45]. The SourceForge repository [122] contains installers for Linux, Mac, and Windows, as well as the complete sources for the most recent Reduce version. Just click on the "Files" tab. Linux users whose distribution provides a recent version of Reduce should install from the distribution's repository (instead of SourceForge) using their distribution's package manager.

If you are ready to recompile Reduce from its source code, please consider the text [106] or the `README.txt` file in the main folder of the source directory for instructions on how to do it in different operating systems. You will need a development environment with a C++ compiler and various development libraries which are provided in different ways by different operating systems.

Please note that installers of Reduce older than 2016 might not have CDE or CDIFF included. The use of such versions is discouraged, although in principle there should be no big problems in recompiling the source files of the two packages for older versions.

A.2 Getting Started with Reduce

Reduce is a computer algebra system, similar to Maple, Mathematica, MuPad (which is now included in Matlab), and Maxima. Reduce can be used in different ways, as all the above programs:

J. Krasil'shchik et al., *The Symbolic Computation of Integrability Structures for Partial Differential Equations*, Texts and Monographs in Symbolic Computation,
https://doi.org/10.1007/978-3-319-71655-8

interactive way: after starting Reduce the user writes commands at a prompt and then requires the execution usually by pressing the "Enter" key. Reduce yields an output and then the prompt is ready again. At this point we need to stress that the command prompt can be presented in two ways:

with GUI: a graphical user interface contains the prompt and is surrounded by menus that help the user in various ways. The GUI usually displays mathematical output in a nice way, and one can re-edit past commands and re-input them again. In Reduce the way to have a GUI is to start the CSL version (see below).

without GUI: the Reduce window is just a text console, without menus and fancy graphical output.

noninteractive way: Reduce is started with a file name as an argument; the file with that name is read, and its lines are executed in a sequential way as Reduce commands. It is also possible to start Reduce, with or without GUI, and then load a program file. This is an interesting possibility as you will monitor the execution of the commands of your program and realize if there are any problems or mistakes.

The noninteractive way is recommended for scientific computations, as the state of Reduce (this means the value of variables and the definition of new commands) or, more generally, the state of a CAS is difficult to control if many complex commands are executed in a typically nontrivial computation. Moreover, a GUI is usually not needed for scientific computations. Indeed, a GUI can make the state of a CAS less clear, as the user can modify and execute a command in previous parts of the worksheet.

The basic capabilities of Reduce are described in the first section of the manual, and it is not worth to repeat them here. Reduce's manual can be found as a PDF file inside the subfolder `doc/manual` of the main directory of your Reduce installation or online at [45]. A good additional beginner's reference on Reduce is the book [87].

Reduce's language is `Rlisp`. `Rlisp` is a language that is similar to `Algol-60`, and to its more famous descendant `Pascal`. The language `Rlisp` is programmed in `Standard Lisp`, which is a dialect of the `Lisp` family of programming languages.[1] Standard `Lisp` was defined in [91] in order to have a definition of the `Lisp` language that would increase portability. In the end `Rlisp` expands to `Standard Lisp`, which is then evaluated by the `Lisp` system at the lowest layer of Reduce; however, writing programs in `Lisp` would be harder than writing programs in `Rlisp` for the majority of users.

Reduce has two underlying `Lisp` systems : CSL and PSL. It is beyond the aims of this book to explain the history and differences between the two systems. However, tests made on CDE show that CSL is generally faster than PSL on our typical problems. The example programs which accompany the book have all been tested using CSL, even if they should work without problems with PSL. Of course, both CSL and PSL are included in Reduce's binary distribution and source code.

[1] `Standard Lisp` was established especially for Reduce, and it is almost completely a subset of `Common Lisp`

In Linux, in order to start CSL, you should run the command `redcsl`, and if you wish to start PSL, the command is `redpsl`. The default behavior for CSL is to start the GUI; run the command `redcsl -w` if you do not want the GUI.

Reduce has two modes of interpreting commands:

algebraic mode: in this mode, Reduce commands can only handle specific data structures, which are said to be *algebraic*, namely, identifiers (or names of variables), algebraic expressions, and algebraic lists. Algebraic expressions are always reduced to their canonical form (when possible) before each calculation. By the way, this explains the name Reduce; other CAS do not share this feature. In order to switch to the algebraic mode, issue the command `algebraic;` at Reduce's terminal; the algebraic mode is the default mode.

symbolic mode: in this mode, Reduce can use any `Lisp` data. `Lisp` data is basically an atom (e.g., the name of a variable) or a list. If you wish to switch to symbolic mode, input the command `symbolic;` at Reduce's terminal

The packages `CDE` and `CDIFF` have been written in symbolic-mode `Rlisp`. Generally speaking, the maximal freedom of expression and speed of execution in Reduce is always achieved at the symbolic-mode level. A normal `CDE` usage would not require using the symbolic mode; all example programs of the book are written in algebraic mode. The readers who are interested in Reduce symbolic mode are kindly invited to have a look at [91, 100, 106]. A good book on Reduce symbolic mode is [19]. Other less complete references are [92, 121].

A Reduce program that makes use of `CDE` can be written with any text editor, even if an editor for writing computer programs fits better to the task. Examples of such editors are given below. It is customary to use the extension `.red` for Reduce programs, like `program.red`. If you wish to run your program, just run the Reduce executable. After starting Reduce, you would see something like

```
Reduce (Free CSL version), 01-Oct-16 ...

1:
```

The working directory of Reduce is the directory where you started it. If the file `program.red` is in the working directory, you can write `in "program.red";` at the prompt "`1:`". If the file `program.red` *is not* in the working directory, you should indicate the full path of the program, and this depends on your system. In Linux, assuming that you are the user `user` and your program is in the subdirectory `Reduce/ computations` of your home directory, you have something like

```
in "/home/user/Reduce/computations/program.red";
```

In Windows, assuming that you are the user `user` and your program is in the subdirectory `Reduce\computations` of the Desktop folder, you would write

```
in "C:\Users\user\Desktop\Reduce
\computations\program.red";
```

If you start Reduce with the GUI, you can use the leftmost menu item `File>Open` in order to avoid to write down the whole path.

The most convenient way to writing Reduce programs is to use an editor for programming languages. It is better to avoid word processors like Microsoft Word or OpenOffice.org as they tend to modify text files by adding some further binary or text data. There are many choices of editors in any operating system. `Emacs` is very stable and efficient; it has specific Reduce support (it is an editing mode called `reduce-ide` [151]); it runs in all operating systems, in particular Linux, Windows, and Mac. Another program which has Reduce support is GNU TeXmacs (see http://www.texmacs.org). In Windows, an alternative suggested text editor is `notepad++` (see https://notepad-plus-plus.org/). Similar tools in Linux are `kwrite`, `gedit`, and `Vim`.

A.3 Getting Started with CDE

In this section, we focus on the technical aspects of `CDE` usage which are not covered in the rest of the book.

All example programs that we discuss in the book can be found at the web page of the book http://gdeq.org/Symbolic_Book and inside your Reduce installation, in the subfolder `packages/cde`. There are some conventions that we adopted when writing the examples. Program files have the following names:

```
equationname_typeofcomputation.red
```

where `equationname` stands for the shortened name of the equation (e.g. Korteweg-de Vries is always indicated by `kdv`) and `typeofcomputation` stands for the type of geometric object which is computed with the given file, for example, symmetries, Hamiltonian operators, etc. This string also includes a version number. More specific information, like the date and more details on the computation done in each version, are included as comment lines at the very beginning of each file.

Please note that the examples are shipped together with their respective results and that they can be run all at once by the shell script `cdiff.sh` (written in `Bash` and packaged with the examples) to test if the system is working properly and results are the same as obtained previously. This can be good if, for example, you introduced new procedures that modify the behavior of `CDE` and you want to test them.

If you use a generic editor, as soon as you are finished writing a program, you may run it from within Reduce by following the instructions in the previous section.

In `Emacs` with Reduce IDE [151] issuing the command `M-x run-reduce` (or choosing the menu item `Run REDUCE>Run REDUCE`) will split the window in two halves and start Reduce in the bottom half. Then you may load the program file that you were editing (suppose that its name is `program.red`) by issuing `in "program.red";` at the Reduce prompt. In fact, `Emacs` lets Reduce assume as its working directory the directory of the file that you were editing.

The results of a computation consist of the values of one or more unknown. Suppose that the unknown's name is sym, and assume that, after a computation, you wish to save the values of sym, possibly for future use from within Reduce. Issue the following Reduce commands (of course, after you finish your computations!):

```
off nat;
out "file_res.red";
sym:=sym;
shut "file_res.red";
on nat;
```

All commands must be terminated with `;`. Let us discuss the effect of the above commands step-by-step:

off nat: yields a text-only output which is suitable for reloading in another Reduce program.

out "file_res.red": opens the file `file_res.red` and erases any previous content in it (you are warned!); moreover, any output is redirected from the terminal to the above file.

sym:=sym: has the effect of writing the left-hand sym unevaluated and writing the *value* of sym to the right-hand side. This is a way to write out your result.

shut "file_res.red": closes the file `file_res.red`; all output between the out and shut command is saved into the above file. The output is redirected into the terminal and is no longer saved.

on nat: restores the previous output behavior.

If you wish to translate your results in LaTeX, you might wish to load the package `tri`, LaTeX-Reduce interface. Just issue the commands

```
load_package tri;
on tex;
```

and all terminal output will be in LaTeX instead of the Reduce text or graphical output format.

Working remotely with Reduce is not difficult, and it is highly recommended for big computations that a server can run more efficiently and without interruptions. There are several methods to do that. We assume that a remote server is running, has Reduce, is connected to the Internet, and is accessible through the SSH ("secure shell") protocol. If the server name is, for example, sophuslie.org, a user can connect by the command[2]

```
ssh username@sophuslie.org
```

The user is then brought into a console where he/she can run Reduce and his/her programs. The programs can be transferred in the server by the SFTP or the SCP protocol. Usually, file transfer programs that use one of the above protocols are provided together with ssh.

[2]There are many SSH programs around for all operating systems, for example, PuTTY or WinSCP for Windows.

If the process must run for a long time, it is inconvenient for the user to stay connected. Ideally, the user would like to launch the process, log out, then log in after some hours (or days), and check the results. Unfortunately, if you just follow the above steps, the operating system will kill any process owned by the user after the user logout. There are several ways to solve this problem.

A simple solution is GNU `screen`. This program allows the user to run many background processes, each one with its name. Initiate a `screen` session:

```
screen
```

You will be brought to a command prompt. Now, if you run any program, including Reduce, this will be in a sense owned by `screen`. If you issue the command `ctrl-a d` (press simultaneously the "ctrl" and the "a" keys, then release them, and press "d"), you will "detach" from the `screen` session. You are free to log out and log in; your computation within the `screen` session will keep running until you will "reattach" to the `screen` session by the command

```
screen -r
```

At the time of writing, a nice tutorial can be found here: http://jmcpherson.org/screen.html.

A similar program that the user might wish to consider is `tmux`. A more elaborated solution is using VNC; we will not go into details about it.

There is the possibility to use the `Emacs` editor as a server which will "own" your computation and keep it running while you are not logged in. Follow the steps below:

(1) log in to the remote server with `ssh`;
(2) start `Emacs` as a daemon on the server by the command `emacs -daemon` (only from version 23.1!);
(3) run `emacsclient -c file.red`. That program will connect to the `Emacs` daemon and open the requested file.
(4) run Reduce (if you installed the reduce IDE everything is easier, otherwise you should open a shell within emacs and issue the command `reduce`);
(5) exit `emacsclient` normally (C-x C-c). This will not kill the daemon, that will keep your computation running until the end.
(6) log in again when you wish to check the computation.

Bibliography

1. Ablowitz, M.J., Clarkson, P.A.: Solitons, Nonlinear Evolution Equations and Inverse Scattering. London Mathematical Society Lecture Note Series, vol. 149. Cambridge University Press, Cambridge (1991)
2. Anderson, I.: Integrable Systems Tools, Maple package available at http://digitalcommons.usu.edu/dg/ (2017)
3. Anderson, I.M., Kamran, N.: Conservation laws and the variational bicomplex for second-order scalar hyperbolic equations in the plane. Acta Appl. Math. **41**(1), 135–144 (1995)
4. Andrews, G.E., Askey, R., Roy, R.: Special Functions. Encyclopedia of Mathematics and its Applications, vol. 71. Cambridge University Press, Cambridge (1999). ISBN:978-0-521-62321-6
5. Apel, J., Hemmecke, R.: Detecting unnecessary reductions in an involutive basis computation. J. Symb. Comput. **40**, 1131–1149 (2005)
6. Baldwin, D., Hereman, W.: A symbolic algorithm for computing recursion operators of nonlinear partial differential equations. Int. J. Comput. Math. **87**(5), 1094–1119 (2010)
7. Barakat, A., De Sole, A., Kac, V.G.: Poisson vertex algebras in the theory of Hamiltonian equations. Jpn. J. Math. **4**, 141–252 (2009)
8. Barakat, M.: Jets. A Maple-package for formal differential geometry. In: Computer Algebra in Scientific Computing, pp. 1–12. Springer, Berlin/Heidelberg (2001)
9. Baran, H., Krasil′shchik, I.S., Morozov, O.I., Vojčák, P.: Higher symmetries of cotangent coverings for Lax-integrable multi-dimensional partial differential equations and lagrangian deformations. In: Konopelchenko, B.G., et al. (eds.) Physics and Mathematics of Nonlinear Phenomena 2013. Journal of Physics: Conference Series, vol. 482, p. 012002 (2014). arXiv:1309.7435
10. Baran, H., Krasil′shchik, I.S., Morozov, O.I., Vojčák, P.: Symmetry reductions and exact solutions of Lax integrable 3-dimensional systems. J. Nonlinear Math. Phys. **21**(4), 643–671 (2014). arXiv:1407.0246
11. Baran, H., Krasil′shchik, I.S., Morozov, O.I., Vojčák, P.: Five-dimensional Lax-integrable equation, its reductions and recursion operator. Lobachevskii J. Math. **36**(3), 225–233 (2015)
12. Baran, H., Krasil′shchik, I.S., Morozov, O.I., Vojčák, P.: Integrability properties of some equations obtained by symmetry reductions. J. Nonlinear Math. Phys. **22**(2), 210–232 (2015)
13. Baran, H., Krasil′shchik, I.S., Morozov, O.I., Vojčák, P.: Coverings over Lax integrable equations and their nonlocal symmetries. Theor. Math. Phys. **188**(3), 1273–1295 (2016). arXiv:1507.00897

J. Krasil'shchik et al., *The Symbolic Computation of Integrability Structures for Partial Differential Equations*, Texts and Monographs in Symbolic Computation,
https://doi.org/10.1007/978-3-319-71655-8

14. Baran, H., Krasil′shchik, I.S., Morozov, O.I., Vojčák, P.: Nonlocal symmetries of Lax integrable equations: a comparative study. Submitted to Theor. Math. Phys. (2016). arXiv:1611.04938
15. Baran, H., Marvan, M.: Jets. A software for differential calculus on jet spaces and diffieties. Silesian University in Opava. First version 2003; revised version 2010. Available at: http://jets.math.slu.cz/
16. Baran, H., Marvan, M.: On integrability of Weingarten surfaces: a forgotten class. J. Phys. A **42**, 404007 (2009)
17. Blaszak, M., Szablikowski, B.M.: Classical r-matrix theory of dispersionless systems: Ii. $(2+1)$-dimension theory. J. Phys. A **35**, 10345 (2002). arXiv:nlin/0211018
18. Bocharov, A.V., Chetverikov, V.N., Duzhin, S.V., Khor′kova, N.G., Krasil′shchik, I.S., Samokhin, A.V., Torkhov, Y.N., Verbovetsky, A.M., Vinogradov, A.M.: In: Krasil′shchik, I.S., Vinogradov, A.M. (eds.) Symmetries and Conservation Laws for Differential Equations of Mathematical Physics. Monograph. American Mathematical Society, Providence (1999)
19. Brackx, F., Constales, D.: Computer Algebra With Lisp and Reduce: An Introduction to Computer-Aided Pure Mathematics. Springer, Dordrecht (1991)
20. Camassa, R., Holm, D.D.: An integrable shallow water equation with peaked solitons. Phys. Rev. Lett. **71**, 1661–1664 (1993)
21. Casati, M.: Higher order dispersive deformations of multidimensional Poisson brackets of hydrodynamic type. (2017). arXiv:1710.08175
22. Casati, M., Valeri, D.: MasterPVA and WAlg: mathematica packages for poisson vertex algebras and classical affine $\mathscr{W}$-algebras. https://arxiv.org/abs/1603.05028 (submitted, 2016)
23. Coley, A., Levi, D., Milson, R., Rogers, C., Winternitz, P. (eds.): Bäcklund and Darboux Transformations. The Geometry of Solitons. CRM Proceedings and Lecture Notes, vol. 29. American Mathematical Society, Providence (2001)
24. Dorfman, I.Y.: Dirac Structures and Integrability of Nonlinear Evolution Equations. Wiley, Chichester/New York (1993)
25. Doubrov, B., Ferapontov, E.V.: On the integrability of symplectic Monge–Ampère equations. J. Geom. Phys. **60**, 1604–1616 (2010). arXiv:0910.3407
26. Dubrovin, B.A., Krichever, I.M., Novikov, S.P.: Integrable systems. I. In: Dynamical Systems IV. Encyclopaedia of Mathematical Sciences, vol. 4, 2nd edn., pp. 173–280. Springer, Berlin (2001)
27. Dubrovin, B.A., Novikov, S.P.: Poisson brackets of hydrodynamic type. Soviet Math. Dokl. **30**, 651–654 (1984)
28. Dubrovin, B.A.: Geometry of 2d topological field theories. Lect. Notes Math. **1620**, 120–348 (1996)
29. Dunajski, M., Kryński, W.: Einstein-Weyl geometry, dispersionless Hirota equation and Veronese webs. Math. Proc. Camb. Philos. Soc. **157**(1), 139–150 (2014)
30. Ferapontov, E.V.: Surfaces in 3-space possessing nontrivial deformations which preserve the shape operator. In: Integrable Systems in Differential Geometry, Tokyo, July 2000, Contemporary Mathematics, vol. 508, pp. 145–159. AMS, Providence (2002). arXiv:math/0107122
31. Ferapontov, E.V., Galvao, C.A.P., Mokhov, O., Nutku, Y.: Bi-hamiltonian structure of equations of associativity in 2-d topological field theory. Commun. Math. Phys. **186**, 649–669 (1997)
32. Ferapontov, E.V., Lorenzoni, P., Savoldi, A.: Hamiltonian operators of Dubrovin–Novikov type in $2d$. Lett. Math. Phys. **105**(3), 341–377 (2014). arXiv:1312.0475
33. Ferapontov, E.V., Pavlov, M.V., Vitolo, R.F.: Projective-geometric aspects of homogeneous third-order hamiltonian operators. J. Geom. Phys. **85**, 16–28 (2014). https://doi.org/10.1016/j.geomphys.2014.05.027
34. Ferapontov, E.V., Pavlov, M.V., Vitolo, R.F.: Towards the classification of homogeneous third-order Hamiltonian operators. Int. Math. Res. Not. **22**, 6829–6855 (2016)
35. Ferapontov, E.V., Pavlov, M.V., Vitolo, R.F.: Systems of conservation laws with third-order Hamiltonian structures', to appear in Lett. Math. Phys. (2018)

36. Fordy, A.P.: A historical introduction to solitons and Bäcklund transformations. In: Fordy, A.P., Wood, J.C. (eds.) Harmonic Maps and Integrable Systems. Aspects of Mathematics, vol. 23, pp. 7–28. Springer Fachmedien Wiesbaden, Braunschweig/Wiesbaden (1994)
37. Fuchssteiner, B.: Mastersymmetries, higher order time-dependent symmetries and conserved densities of nonlinear evolution equations. Prog. Theor. Phys. **70**(6), 1508–1522 (1983)
38. Fuchssteiner, B., Fokas, A.S.: Symplectic structures, their Bäcklund transformations and hereditary symmetries. Phys. D **4**(1), 47–66 (1981)
39. Gardner, C.S., Greene, J.M., Kruskal, M.D., Miura, R.M.: Method for solving the Korteweg-de Vries equation. Phys. Rev. Lett. **19**, 1095–1097 (1967)
40. Getzler, E.: A Darboux theorem for Hamiltonian operators in the formal calculus of variations. Duke Math. J. **111**, 535–560 (2002)
41. Gibbons, J., Tsarev, S.P.: Reductions of the Benney equations. Phys. Lett. A **211**, 19–24 (1996)
42. Göktas, Ü., Hereman, W.: Symbolic computation of conserved densities for systems of nonlinear evolution equations. J. Symb. Comput. **24**(5), 591–621 (1997)
43. Golovko, V.A., Kersten, P.H.M., Krasil′shchik, I.S., Verbovetsky, A.M.: On integrability of the Camassa-Holm equation and its invariants. Acta Appl. Math. **101**, 59–83 (2008)
44. Gumral, H., Nutku, Y.: Bi-hamiltonian structures of D-Boussinesq and Benney–Lax equations. J. Phys. A **27**:193–200 (1994)
45. Hearn, A.C.: Reduce. http://reduce-algebra.sourceforge.net/, version 3.8 edition, 2004. Computer algebra system, currently in development after that it has been released in 2008 as free software at Sourceforge. The manual is available at the website
46. Hereman, W., Adams, P.J., Eklund, H.L., Hickman, M.S., Herbst, B.M.: Direct methods and symbolic software for conservation laws of nonlinear equations. In: Yan, Z. (ed.) Advances in Nonlinear Waves and Symbolic Computation, chap. 2, pp. 19–79. Nova Science Publishers, New York (2009)
47. Husain, V.: Self-dual gravity as a two dimensional theory and conservation laws. Classical Quantum Gravity **11**, 927–937 (1994). arXiv:gr-qc/9310003
48. Ibragimov, N.: Transformation Groups Applied to Mathematical Physics. Reidel, Dordrecht (1985)
49. Igonin, S.: Coverings and fundamental algebras for partial differential equations. J. Geom. Phys. **56**, 939–998 (2006)
50. Igonin, S., Krasil′shchik, J.: On one-parametric families of Bäcklund transformations. In: Morimoto, T., Sato, H., Yamaguchi, K. (eds.) Lie Groups, Geometric Structures and Differential Equations—One Hundred Years After Sophus Lie. Advanced Studies in Pure Mathematics, vol. 37, pp. 99–114. Mathematical Society of Japan, Tokyo (2002)
51. Igonin, S., Verbovetsky, A., Vitolo, R.: Variational multivectors and brackets in the geometry of jet spaces. In: Symmetry in Nonlinear Mathematical Physics. Part 3, pp. 1335–1342. Institute of Mathematics of NAS of Ukraine, Kiev (2003)
52. Janet, M.: Leçons sur les Systèmes d'Équations aux Dérivées Partielles. Gauthier-Villars, Paris (1929)
53. Kadomtsev, B.B., Petviashvili, V.I.: On the stability of solitary waves in weakly dispersive media. Sov. Phys. Dokl. **15**, 539–541 (1970)
54. Kersten, P.: Supersymmetries and recursion operator for $N = 2$ supersymmetric KdV-equation. Sūrikaisekikenkyūsho Kōkyūroku **1150**, 153–161 (2000)
55. Kersten, P., Krasil′shchik, I., Verbovetsky, A.: An extensive study of the $N = 1$ supersymmetric KdV equation. Memorandum 1656, Faculty of Mathematical Sciences, University of Twente, The Netherlands (2002)
56. Kersten, P., Krasil′shchik, I., Verbovetsky, A.: Nonlocal constructions in the geometry of PDE. In: Symmetry in Nonlinear Mathematical Physics. Part 1, pp. 412–423. Institute of Mathematics of NAS of Ukraine, Kiev (2003)
57. Kersten, P., Krasil′shchik, I., Verbovetsky, A.: Hamiltonian operators and ℓ^*-coverings. J. Geom. Phys. **50**, 273–302 (2004)

58. Kersten, P., Krasil′shchik, I., Verbovetsky, A.: The Monge–Ampère equation: Hamiltonian and symplectic structures, recursions, and hierarchies. Memorandum 1727, Faculty of Mathematical Sciences, University of Twente, The Netherlands (2004)
59. Kersten, P., Krasil′shchik, I., Verbovetsky, A.: (Non)local Hamiltonian and symplectic structures, recursions and hierarchies: a new approach and applications to the $N = 1$ supersymmetric KdV equation. J. Phys. A **37**, 5003–5019 (2004)
60. Kersten, P., Krasil′shchik, I., Verbovetsky, A.: On the integrability conditions for some structures related to evolution differential equations. Acta Appl. Math. **83**, 167–173 (2004)
61. Kersten, P., Krasil′shchik, I., Verbovetsky, A.: A geometric study of the dispersionless Boussinesq type equation. Acta Appl. Math. **90**, 143–178 (2006)
62. Kersten, P., Krasil′shchik, I., Verbovetsky, A., Vitolo, R.: On integrable structures for a generalized Monge–Ampère equation. Theor. Math. Phys. **128**(2), 600–615 (2012)
63. Kersten, P., Krasil′shchik, J.: Complete integrability of the coupled KdV-mKdV system. In: Morimoto, T., Sato, H., Yamaguchi, K. (eds.) Lie Groups, Geometric Structures and Differential Equations—One Hundred Years After Sophus Lie. Advanced Studies in Pure Mathematics, vol. 37, pp. 151–171. Mathematical Society of Japan, Tokyo (2002)
64. Khor′'kova, N.G.: On the $\mathscr{C}$-spectral sequence of differential equations. Differ. Geom. Appl. **3**(3), 219–243 (1993)
65. Khor′'kova, N.G.: On the $\mathscr{C}$-spectral sequence for systems of evolution equations. Acta Appl. Math. **41**(1–3), 145–152 (1995)
66. Khor′kova, N.G.: Conservation laws and nonlocal symmetries. Math. Notes **44**, 562–568 (1989)
67. Konopelchenko, B.G.: Nonlinear integrable equations. Lecture Notes in Physics, vol. 270. Springer, Berlin/Heidelberg (1987)
68. Konopelchenko, B.G.: Solitons in Multidimensions: Inverse Spectral Transform Method. World Scientific, Singapore (1993)
69. Krasil′shchik, I.: A natural geometric construction underlying a class of Lax pairs. Lobachevskii J. Math. **37**(1), 60–65 (2016). arXiv:1401.0612
70. Krasil′shchik, I.S.: Algebras with flat connections and symmetries of differential equations. In: Komrakov, B.P., Krasil′shchik, I.S., Litvinov, G.L., Sossinsky, A.B. (eds.) Lie Groups and Lie Algebras: Their Representations, Generalizations and Applications, pp. 407–424. Kluwer, Dordrecht/Boston (1998)
71. Krasil′shchik, I.S., Kersten, P.H.M.: Deformations and recursion operators for evolution equations. In: Prastaro, A., Rassias, T.M. (eds.) Geometry in Partial Differential Equations, pp. 114–154. World Scientific, Singapore (1994)
72. Krasil′shchik, I.S., Kersten, P.H.M.: Graded differential equations and their deformations: a computational theory for recursion operators. Acta Appl. Math. **41**, 167–191 (1994)
73. Krasil′shchik, I.S., Kersten, P.H.M.: Graded Frölicher-Nijenhuis brackets and the theory of recursion operators for super differential equations. In: Lychagin, V.V. (ed) The Interplay between Differential Geometry and Differential Equations. American Mathematical Society Translations, vol. 2, pp. 143–164. American Mathematical Society, Providence (1995)
74. Krasil′shchik, I.S., Kersten, P.H.M.: Symmetries and Recursion Operators for Classical and Supersymmetric Differential Equations. Kluwer, Dordrecht/Boston (2000)
75. Krasil′shchik, I.S., Lychagin, V.V., Vinogradov, A.M.: Geometry of Jet Spaces and Nonlinear Partial Differential Equations. Gordon and Breach, New York (1986)
76. Krasil′shchik, I.S., Morozov, O.I., Sergyeyev, A.: Infinitely many nonlocal conservation laws for the ABC equation with $A + B + C \neq 0$. Calc. Var. Partial Differ. Equ. **55**(5), 1–12 (2016). arXiv:1511.09430
77. Krasil′shchik, I.S., Sergyeyev, A.: Integrability of S-deformable surfaces: conservation laws, Hamiltonian structures and more. J. Geom. Phys. **97**, 266–278 (2015). arXiv:1501.07171
78. Krasil′shchik, I.S., Vinogradov, A.M.: Nonlocal trends in the geometry of differential equations: symmetries, conservation laws, and Bäcklund transformations. Acta Appl. Math. **15**, 161–209 (1989)

79. Krasil′shchik, J., Verbovetsky, A., Vitolo, R.: A unified approach to computation of integrable structures. Acta Appl. Math. **120**(1), 199–218 (2012)
80. Krasil′shchik, J., Verbovetsky, A., Vitolo, R.: The Symbolic Computation of Integrability Structures for Partial Differential Equations. Texts and Monographs in Symbolic Computation. Springer (2018). ISBN:978-3-319-71654-1; to appear; see http://gdeq.org/Symbolic_Book for downloading program files that are discussed in the book
81. Krasil′shchik, J., Verbovetsky, A.M.: Homological Methods in Equations of Mathematical Physics. Advanced Texts in Mathematics. Open Education & Sciences, Opava (1998)
82. Krasil′shchik, J., Verbovetsky, A.M.: Geometry of jet spaces and integrable systems. J. Geom. Phys. **61**, 1633–1674 (2011). arXiv:1002.0077
83. Kuperschmidt, B.A.: KP or mKP: Noncommutative Mathematics of Lagrangian, Hamiltonian, and Integrable Systems. Mathematical Surveys and Monographs, vol. 78. AMS, Providence (2000)
84. Leites, D.A.: Introduction to the theory of supermanifolds. Russ. Math. Surv. **35**(1), 1–64 (1980)
85. Leo, M., Leo, R.A., Martina, L., Pirani, F.A.E., Soliani, G.: Non-abelian prolongation and complete integrability. Phys. D **4**(1), 105–112 (1981)
86. Lorenzoni, P., Savoldi, A., Vitolo, R.: Bi-Hamiltonian systems of KdV type. J. Phys. A : Math. Theor. **51**(4) (2018), 045202. http://arxiv.org/abs/1607.07020
87. MacCallum, M.A.H., Wright, F.J.: Algebraic Computing with Reduce. Lecture Notes from the First Brazilian School on Computer Algebra, vol. 1. Clarendon Press, Oxford (1991). ISBN:978-0-19-853443-3
88. Magri, F.: A simple model of the integrable Hamiltonian equation. J. Math. Phys. **19**, 1156–1162 (1978)
89. Maltsev, A.Y., Novikov, S.P.: On the local systems Hamiltonian in the weakly non-local Poisson brackets. Phys. D **156**(1–2), 53–80 (2001)
90. Malykh, A.A., Nutku, Y., Sheftel, M.B.: Partner symmetries of the complex Monge–Ampère equation yield hyper-Kähler metrics without continuous symmetries. J. Phys. A **36**(39), 10023 (2003). arXiv:math-ph/0305037
91. Marti, J., Hearn, A.C., Griss, M.L., Griss, C.: The Standard Lisp Report. Available in the `/doc/primers` subfolder of Reduce installation.
92. Marti, J.: RLISP ’88: An Evolutionary Approach to Program Design and Reuse. World Scientific, Singapore (1993)
93. Martínez Alonso, L., Shabat, A.B.: Energy-dependent potentials revisited: a universal hierarchy of hydrodynamic type. Phys. Lett. A **299**, 359–365 (2002)
94. Martínez Alonso, L., Shabat, A.B.: Hydrodynamic reductions and solutions of a universal hierarchy. Theor. Math. Phys. **140**, 1073–1085 (2004)
95. Marvan, M.: On zero-curvature representations of partial differential equations. In: Differential Geometry and Its Applications. Proceedings of the Conference, Opava, 1992, pp. 103–122. Open Education & Sciences, Opava (1993)
96. Marvan, M.: Another look on recursion operators. In: Differential Geometry and Applications. Proceedings of the Conference, Brno, 1995, pp. 393–402. Masaryk University, Brno (1996)
97. Marvan, M.: Sufficient set of integrability conditions of an orthonomic system. Found. Comput. Math. **9**, 651–674 (2009)
98. Marvan, M.: On the spectral parameter problem. Acta Appl. Math. **109**(1), 239–255 (2010). arXiv:0804.2031
99. Marvan, M., Sergyeyev, A.: Recursion operators for dispersionless integrable systems in any dimension. Inverse Prob. **28**, 025011 (2012). arXiv:nlin/1107.0784
100. Melenk, H.: Reduce symbolic mode primer. Available in the `/doc/primers` subfolder of Reduce installation
101. Meshkov, A.G.: Tools for symmetry analysis of PDEs. Differ. Equ. Control Proc. **1** (2002). http://www.math.spbu.ru/diffjournal/

102. Morozov, O.I.: Recursion operators and nonlocal symmetries for integrable rmdKP and rdDym equations (2012). arXiv:nlin/1202.2308
103. Morozov, O.I.: A recursion operator for the Universal Hierarchy equation via Cartan's method of equivalence. Cent. Eur. J. Math. **12**(2), 271–283 (2014)
104. Morozov, O.I., Sergyeyev, A.: The four-dimensional Martínez Alonso–Shabat equation: reductions and nonlocal symmetries. J. Geom. Phys. **85**(11), 40–45 (2014)
105. Neyzi, F., Nutku, Y., Sheftel, M.B.: Multi-hamiltonian structure of Plebanski's second heavenly equation. J. Phys. A **38**, 8473 (2005)
106. Norman, A.C., Vitolo, R.: Inside Reduce. Part of the official Reduce documentation, included in the source code of Reduce. See also http://reduce-algebra.sourceforge.net/lisp-docs/insidereduce.pdf (2014)
107. Novikov, S.P., Manakov, S.V., Pitaevskii, L.P., Zakharov, V.E.: Theory of Solitons. Plenum Press, New York (1984)
108. Nucci, M.C.: Interactive reduce programs for calculating classical, non-classical, and approximate symmetries of differential equations. In: Computational and Applied Mathematics II. Differential Equations, pp. 345–350. Elsevier, Amsterdam (1992)
109. Nucci, M.C.: Interactive reduce programs for calculating Lie point, non-classical, Lie-Bäcklund, and approximate symmetries of differential equations: manual and floppy disk. In: CRC Handbook of Lie Group Analysis of Differential Equations, vol. 3, pp. 415–481. CRC Press, Boca Raton (1996)
110. Oliveri, F.: ReLie, Reduce software and user guide. Technical report, Università degli Studi di Messina (2015). http://mat521.unime.it/oliveri/
111. Ovsienko, V.: Bi-Hamiltonian nature of the equation $u_{tx} = u_{xy}\, u_y - u_{yy}\, u_x$. Adv. Pure Appl. Math. **1**, 7–17 (2010)
112. Pavlov, M.V.: Integrable hydrodynamic chains. J. Math. Phys. **44**, 4134–4156 (2003)
113. Pavlov, M.V., Chang, J.H., Chen, Y.T.: Integrability of the Manakov–Santini hierarchy. arXiv:0910.2400
114. Pavlov, M.V., Vitolo, R.F.: On the bi-Hamiltonian geometry of the WDVV equations. Lett. Math. Phys. **105**(8), 1135–1163 (2015)
115. Plebanski, J.F.: Some solutions of complex Einstein equations. J. Math. Phys. **16**(12), 2395–2402 (1975)
116. Pommaret, J.-F.: Systems of Partial Differential Equations and Lie Pseudogroups. Gordon and Breach, New York (1978)
117. Poole, D., Hereman, W.: Symbolic computation of conservation laws for nonlinear partial differential equations in multiple space dimensions. J. Symb. Comput **46**(12), 1355–1377 (2011)
118. Popovych, R.O., Samoilenko, A.M.: Local conservation laws of second-order evolution equations. J. Phys. A **41**, 362002 (2008). arXiv:0806.2765
119. Popovych, R.O., Sergyeyev, A.: Conservation laws and normal forms of evolution equations. Phys. Lett. A **374**, 2210–2217 (2010). arXiv:1003.1648
120. Post, G.F.: A manual for the package tools 2.1. Technical Report Memorandum 1331, Department of Applied Mathematics, University of Twente (1996). http://gdeq.org/CDIFF
121. Rayna, G.: Reduce: Software for Algebraic Computation. Springer, New York (1987)
122. The Reduce project page at Sourceforge. https://sourceforge.net/projects/reduce-algebra/
123. Riquier, C.: Les Systèmes d'Équations aux Dérivées Partielles. Gauthier-Villars, Paris (1910)
124. Roelofs, G.H.M.: The INTEGRATOR package for Reduce. Technical Report Memorandum 1100, Department of Applied Mathematics, University of Twente (1992). http://gdeq.org/CDIFF
125. Roelofs G.H.M.: The SUPER_VECTORFIELD package for Reduce. Technical Report Memorandum 1099, Department of Applied Mathematics, University of Twente (1992). http://gdeq.org/CDIFF
126. Rogers, C., Schief, W.K.: Bäcklund and Darboux Transformations. Cambridge University Press, Cambridge/New York (2002)

127. Saccomandi, G., Vitolo, R.: On the mathematical and geometrical structure of the determining equations for shear waves in nonlinear isotropic incompressible elastodynamics. J. Math. Phys. **55**, 081502 (2014). arXiv:1408.6177
128. Schwartz, F.: The package SPDE for determining symmetries of partial differential equations. http://reduce-algebra.sourceforge.net/manual/contributed/spde.pdf (1985)
129. Sergyeyev, A.: Locality of symmetries generated by nonhereditary, inhomogeneous, and time-dependent recursion operators: a new application for formal symmetries. Acta Appl. Math. **83**, 95–109 (2004)
130. Sergyeyev, A.: Why nonlocal recursion operators produce local symmetries: new results and applications. J. Phys. A **38**(15), 3397–3407 (2005)
131. Sergyeyev, A.: New integrable (3 + 1)-dimensional systems and contact geometry. Lett. Math. Phys. (2017). https://doi.org/10.1007/s11005-017-1013-4
132. Sergyeyev, A.: Recursion operators for multidimensional integrable PDEs. (2017, Submitted). arXiv:1710.05907
133. Sergyeyev, A.: A simple construction of recursion operators for multidimensional dispersionless integrable systems. J. Math. Anal. Appl. **454**, 468–480 (2017). arXiv:1501.01955
134. Sergyeyev, A., Vitolo, R.: Symmetries and conservation laws for the Karczewska–Rozmej–Rutkowski–Infeld equation. Nonlinear Analysis: Real World Applications. **32**, 1–9 (2016)
135. Sheftel, M.B., Malykh, A.A.: On classification of second-order PDEs possessing partner symmetries. J. Phys. A **42**(39), 395202 (2009). arXiv:0904.2909
136. Sheftel, M.B., Yazıcı, D.: Bi-Hamiltonian representation, symmetries and integrals of mixed heavenly and Husain systems. J. Nonlinear Math. Phys. **7**(4), 453–484 (2010). arXiv:0904.3981
137. Sheftel, M.B., Yazıcı, D., Malykh, A.A.: Recursion operators and bi-Hamiltonian structure of the general heavenly equation. J. Geom. Phys. **116**, 124–139 (2017). arXiv:1510.03666
138. Vinogradov, A.M.: The $\mathscr{C}$-spectral sequence, Lagrangian formalism, and conservation laws. I. The linear theory. II. The nonlinear theory. J. Math. Anal. Appl. **100**, 1–129 (1984)
139. Vinogradov, A.M.: Cohomological Analysis of Partial Differential Equations and Secondary Calculus. American Mathematical Society, Providence (2001)
140. Vinogradov, A.M., Krasil′shchik, I.S.: A method of computing higher symmetries of nonlinear evolution equations and nonlocal symmetries. Sov. Math. Dokl. **22**, 235–239 (1980)
141. Vitolo, R.: CDIFF: a reduce package for computations in geometry of differential equations. User manual available at http://gdeq.org/CDIFF (2011)
142. Vitolo, R.: CDE: a reduce package for integrability of PDEs. Included in Reduce. Example programs are available in the `/packages/cde/examples` of Reduce installation or at the web page http://gdeq.org/CDE (2014)
143. Wahlquist, H.D., Estabrook, F.B.: Prolongation structures of nonlinear evolution equations. J. Math. Phys. **16**, 1–7 (1975)
144. Wahlquist, H.D., Estabrook, F.B.: Prolongation structures of nonlinear evolution equations. II. J. Math. Phys. **17**, 1293–1297 (1976)
145. Wang, J.P.: Symmetries and Conservation Laws of Evolution Equations. PhD thesis, Vrije Universiteit/Thomas Stieltjes Institute, Amsterdam (1998)
146. Wang, J.P.: A list of $1 + 1$ dimensional integrable equations and their properties. J. Nonlinear Math. Phys. **9**, 213–233 (2002)
147. Wolf, T.: An efficiency improved program LIEPDE for determining Lie-symmetries of PDEs. In: Proceedings of Modern Group Analysis: Advanced Analytical and Computational Methods in Mathematical Physics. Kluwer (1993)
148. Wolf, T.: A comparison of four approaches to the calculation of conservation laws. Eur. J. Appl. Math. **13**(2), 129–152 (2002)
149. Wolf, T., Brand, A.: CRACK, user guide, examples and documentation. http://lie.math.brocku.ca/Crack_demo.html
150. Wolf, T., Brand, A.: Investigating des with crack and related programs. SIGSAM Bull. Spec. Issue 1–8 (1995)

151. Wright, F.J.: Reduce-IDE, an integrated development environment for Reduce in Emacs. http://reduce-algebra.sourceforge.net/reduce-ide/
152. Zabolotskaya, E.A., Khokhlov, R.V.: Quasi-plane waves in the nonlinear acoustics of confined beams. Sov. Phys. Acoust. **15**, 35–40 (1969)
153. Zakharevich, I.: Nonlinear wave equation, nonlinear Riemann problem, and the twistor transform of Veronese webs. arXiv:math-ph/0006001
154. Zakharov, V.E. (ed.): What is integrability? Springer, Berlin (1991)
155. Zakharov, V.E., Faddeev, L.D.: The Korteweg-de Vries equation is a fully integrable Hamiltonian system. Funct. Anal. Appl. **5**, 280–287 (1971)
156. Zhiber, A.V., Sokolov, V.V.: Exactly integrable hyperbolic equations of Liouville type. Russ. Math. Surv. **56**(1), 61–101 (2001). https://doi.org/10.4213/rm357

List of CDE Procedures

J. Krasil'shchik et al., *The Symbolic Computation of Integrability Structures for Partial Differential Equations*, Texts and Monographs in Symbolic Computation,
https://doi.org/10.1007/978-3-319-71655-8

List of Example Files

The example files can be found at the web page of this book [80] in the geometry of differential equations website: http://gdeq.org/Symbolic_Book.

J. Krasil'shchik et al., *The Symbolic Computation of Integrability Structures for Partial Differential Equations*, Texts and Monographs in Symbolic Computation,
https://doi.org/10.1007/978-3-319-71655-8

List of CDE Global Variables

The variables whose name ends in !* are symbolic-mode variables; all other variables are algebraic-mode variables. Two variables whose names are the same up to !* represent data which is shared between the algebraic and the symbolic mode. The variables whose value is a number are both symbolic- and algebraic-mode variables at the same time.

all_der_id: all even derivatives in identifier notation. 40

all_der_id!*: all even derivatives in identifier notation. 40

all_der_mind!*: all even derivatives in multiindex notation. 40

all_mind!*: all multiindices generated in the current run. 40

all_mind_table!*: table of multiindices for all variables. 40

all_odd_id: all odd derivatives in identifier notation. 42

all_odd_id!*: all odd derivatives in identifier notation. 42

all_odd_mind!*: all odd derivatives in multiindex notation. 42

all_parametric_der: list of even parametric derivatives ordered according with increasing order of derivative. 43

all_parametric_der!*: list of even parametric derivatives ordered according with increasing order of derivative. 43

J. Krasil'shchik et al., *The Symbolic Computation of Integrability Structures for Partial Differential Equations*, Texts and Monographs in Symbolic Computation,
https://doi.org/10.1007/978-3-319-71655-8

all_parametric_odd: list of odd parametric derivatives ordered according with increasing order of derivative. 44

all_parametric_odd!*: list of odd parametric derivatives ordered according with increasing order of derivative. 44

all_principal_der: list of even principal derivatives ordered according with increasing order of derivative. 43

all_principal_der!*: list of even principal derivatives ordered according with increasing order of derivative. 43

all_principal_odd: list of odd principal derivatives ordered according with increasing order of derivative. 44

all_principal_odd!*: list of odd principal derivatives ordered according with increasing order of derivative. 44

de: right-hand side of the even differential equations. 43

de!*: right-hand side of the even differential equations. 43

deg_dep_var: weight of dependent variables. 72

deg_indep_var: weight of independent variables. 72

deg_odd_var: weight of odd variables. 72

dep_var: list of dependent variables. 39

dep_var!*: list of dependent variables. 40

diffcon_comp_der!*: list of all computed even principal derivatives. 44

diffcon_comp_ext!*: list of all computed `ext` principal derivatives. 44

diffcon_der!*: association list of even principal derivatives and corresponding expression, ordered according with decreasing order of derivatives. 44

diffcon_odd!*: association list of odd principal derivatives and corresponding expression, ordered according with decreasing order of derivatives. 44

diffcon_param_der!*: association list of even principal derivatives and corresponding expressions which are free from principal derivatives. 44

diffcon_param_ext!*: association list of ext principal derivatives and corresponding expressions which are free from principal derivatives. 44

i2m_jetspace!*: association list of even derivative coordinates in identifier and multiindex notation. 40

i2m_jetspace_odd!*: association list of odd derivative coordinates in identifier and multiindex notation. 42

i2o_jetspace!*: association list of even derivative coordinates and their total order of derivative. 40

i2o_jetspace_odd!*: association list of odd derivative coordinates and their total order of derivative. 42

id_tot_der!*: identifier of total derivatives for ext variables, dd is the default. 41

indep_var: list of independent variables. 39

indep_var!*: list of independent variables. 40

letop: auxiliary function depending on independent variables signalling "jet space too small". 41

m2o_jetspace!*: association list of multiindices and their length. 40

n_all_ext: number of all odd-ext variables. 42

n_indep_var: number of independent variables. 40

odd_var: list of odd variables. 42, 44

odd_var!*: list of odd variables. 42, 44

primary_diffcon_der!*: list of even principal derivatives which appear in total derivatives and whose expression must be computed. 44

primary_diffcon_der_tot!*: list of lists; each list contains even primary differential consequences together with the data which identifies it as a coefficient in a total derivative. 44

primary_diffcon_ext_tot!*: list of lists; each list contains even primary differential consequences together with the data which identifies it as a coefficient in a total derivative. 44

Index

J. Krasil'shchik et al., *The Symbolic Computation of Integrability Structures for Partial Differential Equations*, Texts and Monographs in Symbolic Computation,
https://doi.org/10.1007/978-3-319-71655-8

T

The manufacturer's authorised representative in the EU is Springer Nature Customer Service Centre GmbH, Europaplatz 3, 69115 Heidelberg, Germany. If you have any concerns regarding our products, please contact ProductSafety@springernature.com

Printed and bound by CPI Group (UK) Ltd, Croydon, CR0 4YY
01/10/2026
02225951-0013